MICROELECTRONICS EDUCATION

Microelectronics Education

Proceedings of the 5th European Workshop on
Microelectronics Education,
held in Lausanne, Switzerland, April 15–16, 2004

Edited by

Adrian M. Ionescu
*Swiss Federal Institute of Technology,
Lausanne, Switzerland*

Michel Declercq
*Swiss Federal Institute of Technology,
Lausanne, Switzerland*

Maher Kayal
*Swiss Federal Institute of Technology,
Lausanne, Switzerland*

and

Yusuf Leblebici
*Swiss Federal Institute of Technology,
Lausanne, Switzerland*

KLUWER ACADEMIC PUBLISHERS
DORDRECHT / BOSTON / LONDON

A C.I.P. Catalogue record for this book is available from the Library of Congress.

Additional material to this book can be downloaded from http://extras.springer.com

ISBN 978-1-4020-2072-8
ISBN 978-1-4020-2651-5 (eBook)

Published by Kluwer Academic Publishers,
P.O. Box 17, 3300 AA Dordrecht, The Netherlands.

Sold and distributed in North, Central and South America
by Kluwer Academic Publishers,
101 Philip Drive, Norwell, MA 02061, U.S.A.

In all other countries, sold and distributed
by Kluwer Academic Publishers,
P.O. Box 322, 3300 AH Dordrecht, The Netherlands.

Printed on acid-free paper

CONTENTS

CMOS AND NANOELECTRONICS

ORAL PRESENTATIONS

POSTER PRESENTATIONS

INTERNATIONAL EDUCATION

ORAL PRESENTATIONS

MICROSYSTEMS

ORAL PRESENTATIONS

INDUSTRY RELATIONS

ORAL PRESENTATIONS

MULTIMEDIA AND INNOVATIVE TEACHING METHODS

ORAL PRESENTATIONS

POSTER PRESENTATIONS

PREFACE

The 5th edition of the European Workshop on Microelectronics Education (EWME 2004) was organized on April 15th-16th at the Swiss Federal Institute of Technology Lausanne (EPFL), Switzerland and attracted more than 60 participants from 17 countries. The organizer was the Institute of Microelectronics and Microsystems of the School of Engineering of EPFL.

EPFL (Ecole Polytechnique Fédérale de Lausanne) is one of the two national engineering schools in Switzerland. It delivers Engineer (Diploma) and PhD degrees in 12 engineering fields. The School has a staff of about 2'400 people including professors, scientists and administrative personnel. The number of engineering students is more than 6'000, while PhD students count for more than 900. The Institute of Microelectronics and Microsystems (http://imm.epfl.ch/) is part of the new School of Engineering (http://sti.epfl.ch/) of EPFL. IMM includes the Electronics Laboratories (LEG), the Microsystem Laboratories (LMIS) and the Microelectronic Systems Laboratory (LSM) of EPFL. IMM is a research unit with an advanced multidisciplinary research field vocation and has a staff of about 120 professors, researchers and PhDs.

These EWME 2004 proceedings mirror the main technical contributions in advanced teaching and research that were presented at the EWME 2004 workshop, which included 9 main sessions:

- Opening session including messages from professor Nadine Guillemot, president of EWME steering Committee and professor Michel Declercq, Dean of the School of Engineering of EPFL
- Bioengineering and bio-inspired computing
- Design and System-On-Chip
- CMOS and nanoelectronics
- International education
- Microsystems
- Poster session
- Industry relations
- Multimedia and innovative teaching

It is worth noting that for the first time, EWME structured a specific session with invited contributions on bioengineering and bio-inspired computing highlighting the increasing interest in both research and education in this field. Another particular point of focus was on nanoelectronics evolution and revolution, where many questions arise concerning the *after CMOS* era and the multidisciplary preparation of engineers for nano-scale sciences.

EWME 2004 program included some outstanding invited papers dealing with very hot topics in teaching and research:

- *Micro/Nanoelectronics for Life Systems*, by Professor Steve Kang, University of California, Santa Cruz, USA

- *Towards Bio-Inspired Computiong: the Embryonic Project,* by Professor Daniel Mange, Swiss Federal Institute of Technology Lausanne, Switzerland
- *System-On-Chip Curriculum Challenges*, by Professor Hannu Tenhunen, School of Information Technology, Royal Institute of Technology (KTH), Sweden
- *Teaching Nanoelectronic Devices,* by Professor Jerry Fossum, University of Florida, Gainesville, USA
- *The Future of CMOS Nanoelectronics,* by Dr. Simon Deleonibus, LETI-CEA, Grenoble, France
- *Microfluidics chips for single cell manipulation and electrical analysis,* by Professor Philippe Renaud, Swiss Federal Institute of Technology Lausanne, Switzerland
- *Integrating Mixed Signal IC Design Research into a Project-based Undergraduate Miroelectronics Curriculum,* by Professor John McNeill, Worchester Polytechnic Institute, Massachusetts, USA

The devoted help of the EPFL staff guaranteed the success of the conference organization. We particularly acknowledge the efforts of Marie Halm, Isabelle Buzzi, Séverine Eggli, and Raymond Sutter, and the following PhD students and/or post doctoral fellows: Kirsten Moselund, Kathy Buchheit, Costin Anghel, Nasser Hefyene, Vincent Pott, Serge Ecoffey, Raphaël Fritschi, Marcelo Pisani, Santanu Mahapatra, Yogesh Singh Chauhan, Alexandre Mehdaoui and Nicolas Abelé.

Adrian M. Ionescu, Michel Declercq, Yusuf Leblebici, Maher Kayal
Swiss Federal Institute of Technology, Lausanne, Switzerland

BIOENGINEERING AND BIO-INSPIRED COMPUTING

ORAL PRESENTATION

MICRO/NANOELECTRONICS FOR LIFE SYSTEMS

KANG S-M.S.
University of California, Santa Cruz
2004 European Workshop for Microelectronics Education

Abstract

As the silicon workhorse CMOS moves into sub-100nm regime, it opens many new doors to enhance the quality of life, especially in medical application areas. It is predicted that in the next few decades with advancement of bioinformatics genes will be linked to particular diseases and more individualized medicine will be provided based on genetic variation. Medical doctors are expected to make millions of measurements for each patient using the help of technological advancements and large-scale computation for accurate diagnosis of diseases and treatment with minimal side effect. One of the most recent NSF Engineering Research Centers, Biomimetic Microelectronic Systems (BMES) ERC calls for interdisciplinary research and education among three universities, University of Southern California (lead campus), California Institute of Technology, and University of California at Santa Cruz, to develop implantable silicon chips for retinal prosthesis, neuromuscular prosthesis, and cortical prosthesis to restore vision, movement, and cognition. Chips need to be reliable, of ultra low power, and implantable. NASA's mission for space exploration calls for innovative miniaturized devices for dependable computing, intelligent sensing, visual communications, and fuel economy. In essence we will develop a monolithic or a single package fusion of information technology, biotechnology, and nanotechnology for various applications for wholistic life systems. Thus one of the significant challenges lies in interdisciplinary education of micro/nanoelectronics with deep molecular understanding of systems biology, interface between electrodes and living cells. In the Baskin School of Engineering, we have put our foci on IT, BT, and NT and identified the crucial intersection of these three areas to be in Biomolecular Engineering which includes bioinformatics, nanobiosensors, protein engineering, among others. Several faculty members in Electrical Engineering are working on bioelectronics for nanobiosensors, electrode-living cell interaction, and nanobiophotonics for single molecule level sensor in genome sequencing. In this talk we will highlight some of the exciting research projects and the curriculum of biomolecular engineering and related programs.

5

A.M. Ionescu et al. (eds.), Microelectronics Education, 5.
© 2004 *Kluwer Academic Publishers.*

DESIGN AND SYSTEM-ON-CHIP

ORAL PRESENTATIONS

AN OPEN SYSTEM-ON-CHIP PLATFORM FOR EDUCATION

BOULDIN D-W. AND SRIVASTAVA R-R.
Electrical & Computer Engineering
University of Tennessee
Knoxville, TN 37996-2100
dbouldin@tennessee.edu

1.   INTRODUCTION

Million-gate integrated circuits are increasingly being designed as system-on-chip (SoC) platforms since platform design mitigates the risks involved with integrating a CPU core and other virtual components by a fixed deadline. Using this approach, designers can overcome uncertainties about the quality of the components and their interaction and can produce derivative designs rapidly. The development of a SoC platform is described in this paper. In the process of conducting this team project, students learned not only to reuse existing cores but also the requirements to create high quality cores for reuse. The SoC platform, which uses only open cores that can be obtained by anyone at no charge, can also serve as an "industrial strength" design for students to learn about optimizations at the logic and physical levels. Thus, students can exercise synthesis and place/route tools to explore the power-delay-area solution space of a million-gate design. Having internal visibility of the components at both the source code level and at the physical layout level greatly facilitates their understanding of SoC issues. The SoC platform is being placed in the public domain so that others may contribute to its enhancement.

2.   Project Goals and Core Selection

Today designers of application-integrated circuits are faced with the challenge of creating and verifying the content of million-transistor chips as quickly as possible in order to reduce the time-to-market [1]. It has been estimated that a one-month delay in bringing a product to market can result in a loss of ten percent of the potential revenue [2]. Hence, not all of the transistors on these chips can be customized but instead must be ported from previous designs. These reusable cores or intellectual property (IP) blocks include CPUs (like ARM, PowerPC and LEON), MPEG decompression engines, PCI bus controllers, specialized DSPs, etc. Combining several complex cores using standard cells is much more manageable and quicker than designing millions of transistors one at a time. The myth that characterizes today's IP is that these components are blocks that have well-defined contents and interfaces. However, they are often fuzzy and hence appear more like patches in a quilt, which must be stitched together. The

A.M. Ionescu et al. (eds.), Microelectronics Education, 11–15.
© 2004 *Kluwer Academic Publishers.*

components cannot be assembled blindly and rapidly, but rather must be carefully pieced together to form a working system. Therefore, design for reuse does not come free.

Thus, universities and individuals can and are developing open SoCs to serve as education and research platforms [3]. In our graduate program at the University of Tennessee [4], ECE 652 involves advanced physical level design. Thus, students exercise synthesis and place/route tools to explore the power-delay-area solution space of a million-gate design. Having internal visibility of the components at both the source code level and at the physical layout level greatly facilitates their understanding of SoC issues.

In support of the goals of ECE 652, we designed and implemented a baseline SoC platform targeting the TSMC-0.18 CMOS process. To enhance the students' understanding of SoC issues, we selected only open soft cores that could be obtained for free [5-6] or generated internally. For the CPU, we selected the LEON-2 processor [5], which is Sparc-V8 compatible. As shown in Figure 1, it provides direct memory interfaced PROM, memory mapped I/O, SRAM, SDRAM with variable memory width of 8, 16 or 32 bits. The LEON-2 processor can also include various other features such as two UARTs, interrupt controller, memory controller, and an interface for a coprocessor or floating point unit. A flexible configuration scheme makes it straightforward to add new cores as masters or slaves depending upon their functionality. The LEON-2 processor also has separate data and instruction cache RAMs which can be generated in 1-4 sets each of 1-64Kb depending on the functionality desired. The compiler for the LEON-2 is LECCS (Leon/Erc32 GNU Cross-Compiler System) which is compatible with Sun Solaris / Linux / Windows operating systems. LECCS supports ordinary sequential C/C++ programming or multitasking using RTEMS (Real Time Embedded Micro-controller Systems) kernel.

3.   Platform SoC Design and Verification Flow

The entire ECE 652 class of sixteen graduate students was divided into small groups working independently on cores so it was essential to define some specifications and standards to enable the integration of these into a complete SoC at a later stage. Each core was verified individually via pre-layout simulation, synthesis, place/route and post-layout simulation prior to attempting integration with the LEON-2 or other cores. Thus, we could be assured that adding a new core to our SoC design would not introduce any errors within that core and we need only test for its interaction with the rest of the SoC platform.

The task of integrating these cores into a SoC platform is greatly facilitated by using a common bus protocol to interconnect them. For this purpose, an AMBA– wrapper was created for each core such that it would enable the cores to act as AHB bus masters and APB bus slaves. Although each core had a different data width and operating procedures, it was decided that the data width would be 32 bits and each core would start its work when given a proper control signal. The size of the RAMs added to individual cores was also fixed. At this point our strategy of predefining the style of communication among cores proved most advantageous, as the same wrapper could be used for other cores with minor modifications. Different cores can access on-chip peripherals using AMBA busses and simultaneous bus requests can be handled by defining priorities for each core. A core acting as a bus master can access the bus only when granted permission by the AHB controller. However, it can perform its function once the data has been loaded into its RAM and given the go-ahead signal without the need to continue having bus control.

Various design constraints and incremental synthesis techniques were used to achieve a correct simulation of the LEON-2 processor with other cores. Artisan RAMs were added to the design and, with proper power planning techniques, place and route of the design was completed.

Conclusions and Plans

An open SoC platform, which uses only open cores that can be obtained by anyone at no charge, has been developed, implemented and verified as an "industrial strength" design for students to learn about optimizations at the logic and physical levels. The SoC platform is being placed in the public domain so that others may contribute to its enhancement.

ACKNOWLEDGMENTS

We gratefully acknowledge Synopsys, Mentor Graphics and Cadence Design Systems for support of the laboratory with their integrated circuit design tools.

REFERENCES

[1]   International Technology Roadmap for Semiconductors, http://public.itrs.net/

[2]   Smith, M. J. S., *Application-Specific Integrated Circuits,* Addison-Wesley, Boston, MA, 1997.

[3]   Bolado, M., Castillo, J., Posadas, H., Sanchez, P., Villar, E., Sanchez, C., Blasco, P. and H. Fouren, "Platform Based on Open-Source Cores for Industrial Applications", *Proceedings of the 2004 Design Automation and Test in Europe,* Paris, France, (to appear), 16-20 Feb. 2004.

[4]   Bouldin, D.W, *Microelectronic Systems Courses,* University of Tennessee, http://vlsi1.engr.utk.edu/ece/bouldin_courses/

[5]   Gaisler Research, http://www.gaisler.com/

[6]   Open Cores, http://www.opencores.org/

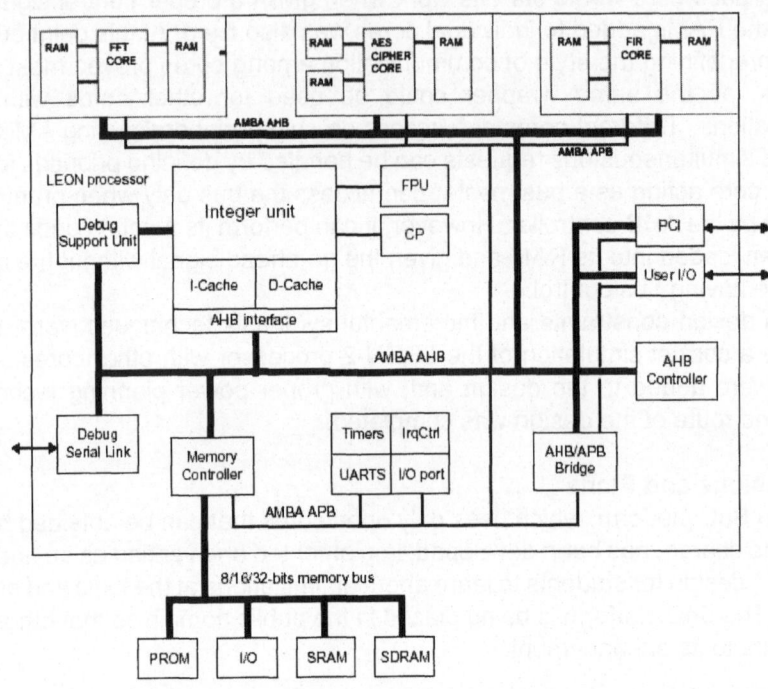

Fig. 1: Block Diagram of the Open SoC Platform.

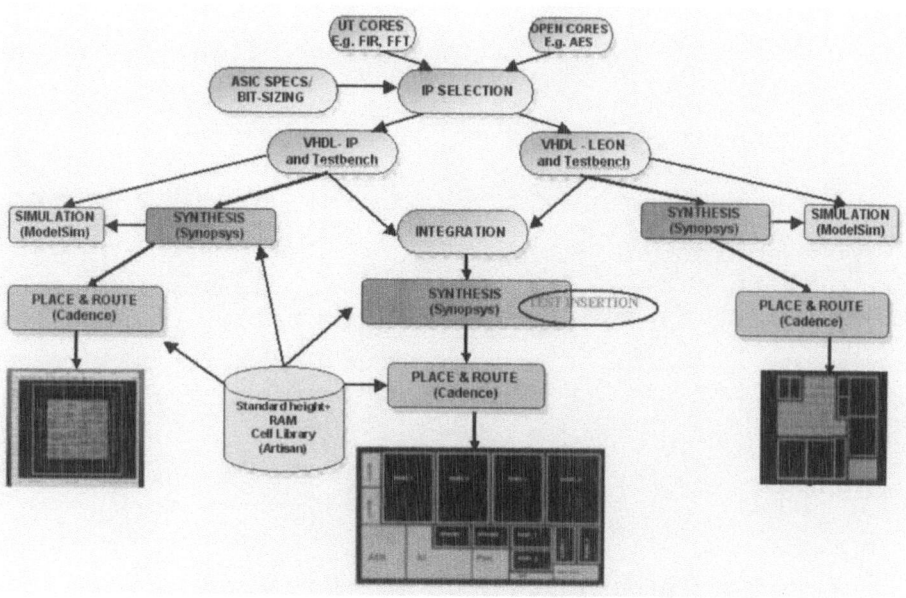

Fig. 2: Platform SoC Design and Verification Flow

MARKET-DRIVEN DESIGN PROJECT IN INTRODUCTORY VLSI DESIGN

BRAUER E-J.
Northern Arizona University, College of Engineering and Technology, PO Box 15600, Flagstaff, AZ USA 86011-5600

1.   INTRODUCTION

The complexity of today's CMOS integrated circuits is truly astounding, with hundreds of millions of transistors performing digital functions, RF circuits operating in the 1 to 10 GHz range, and analog circuits performing very accurate calculations. At the same time, the market pressures are enormous with customers expecting high performance at low cost and low power and competitors rapidly introducing new chips. In this environment, the semiconductor industry requires knowledgeable, capable engineers with expertise in the VLSI area.

This paper addresses the educational issue of introducing senior-level and beginning graduate students in electrical engineering to the area of VLSI design. The case study is EE 482 Introduction to VLSI Design taught in the Electrical Engineering Department, College of Engineering and Technology at Northern Arizona University (NAU). The design project is a market-driven project, which rewards early completion of assignments in connection with learning the fundamental concepts of VLSI.

2.   COURSE OVERVIEW

I have taught this class for the past 6 years at NAU using industry-standard computer-aided design tools from Mentor Graphics Corporation. The class contains the typical mix of homework assignments and hourly exams as well as a unique market-driven design project.

The course is organized to achieve the following objectives. The student is able to
- Design CMOS circuits to meet static, transient, power and area specifications
- Use electronic computer-aided design software to layout and simulate digital CMOS circuits

The focus is on silicon CMOS since this represents 70% of the semiconductor industry and will serve as a foundation for further study in VLSI. The application is digital in this introductory course.

NAU has a 15-week semester with a final exam week. There are 29 lectures in the semester. The class meets weekly for 2, 50-minute periods of lecture and 1, 2-hour period for computer lab to work on the project. The computer lab has open hours for additional lab work as needed by individual students.

A.M. Ionescu et al. (eds.), Microelectronics Education, 17–21.
© 2004 *Kluwer Academic Publishers.*

2.1   EVALUATION OF STUDENT PERFORMANCE

Student performance is evaluated on the basis of weekly homework assignments (10%), class participation and attendance (10%), two hourly exams (15% each), final exam (20%), and the design project (30%). I also require that students complete the online course evaluation, with a penalty of one letter grade for non-completion. I provide a list of 15 to 20 objectives for each exam. Students are permitted an additional sheet of notes for each exam, which allows the students to focus on the concepts rather than memorizing equations.

The final exam is cumulative, covering the entire semester's material with a slight emphasis on the material after the second hourly exam, thus giving students another chance to review material and make connections between the topics covered in the first of the semester with the topics covered later.

The students are required to participate in lecture. I call on specific students to answer questions about the lecture material. This puts the student on the spot a bit but does serve several important functions. One, it is important feedback for me whether the students are grasping the concepts or not. Two, engineers are often required to discuss technical concepts at group meetings or in individual discussion with colleagues so this class is good practice. Three, students stay alert in my class because they may be called on at any moment.

2.2   COURSE CONTENT

The textbook is *CMOS Digital Integrated Circuits: Analysis and Design* by Steve Kang and Yusuf Leblebici, published by McGraw-Hill [1]. The first topics of the course treat the transistor as a switch and discuss the operation of basic CMOS gates such as combinational logic, transmission gates, latches, and flip-flops. This allows the students to get a quick introduction to CMOS circuits and how they function, as well as provides some background for beginning the design project in the first week of classes. The next phase of the course looks at the MOS transistor is some detail in terms of fabrication, layout, Level 1 IV model, and capacitances. The first hourly exam covers all the topics to this point.

The next topics are detailed analysis of the inverter performance: static, transient, area, and power, although the area criterion is covered primarily in the design project. Interconnect is discussed along with the purpose and operation of more complex CMOS circuits, such as complex static logic, superbuffer, dynamic logic, and pass transistors. The second hourly exam covers the topics to this point.

In the last third of the course, I cover high-performance dynamic circuits and semiconductor memories (DRAM, SRAM, Flash). The final exam concludes the course.

The market-driven design project, discussed in detail in the next section, is started the very first week and is continued throughout the course, in parallel with the lecture. The URL for the class is http://jan.ucc.nau.edu/~ejb3/ee482/.

3.   MARKET-DRIVEN DESIGN PROJECT

3.1   DESIGN PROJECT

The design project consists of schematic entry, simulation, layout, and back-annotation of various digital logic circuits in full-custom and standard cell design styles. The full-custom design progresses bottom-up from a 5-input nor gate, 5-bit 2-to-1 multiplexer, full adder, flip-flip, 5-bit full adder, and 5-bit register, to the top level cell of either a modulo counter or up-down counter. See Figure 1 for the up-down counter schematic. The modulo counter is similar. Each student is assigned to either Track A or Track B and completes the specified cells as shown in Table 1.

Students use the schematic-driven layout process with the software tools from Mentor Graphics Corporation: Design Architect for schematic entry, ICGraph for layout, AccuSim for circuit level simulation, and QuickSim for logic level simulation. See Figure 2 for the design flow.

Project	Track A	Track B
Design 1	Schematic: nor5, mux5	Schematic: nor5, mux5
Design 2	Schematic: fa	Schematic: ff
Design 3	Layout: nor5, mux5	Layout: nor5, mux5
Design 4	Layout: fa	Layout: ff
Design 5	Schematic & layout: fa5	Schematic & layout: ff5
Design 6	Std cell: modulo counter	Std cell: up-down counter
Design 7	Schematic & layout: modulo counter	Schematic & layout: up-down counter
Design 8	Final report	Final report

Table 1. Cells required for design project

For the standard cell design, the students are provided with the logic of the modulo or up-down counter and runs a logic simulator (QuickSim) with a provided macro file to verify the correct operation of the circuit. This illustrates that various simulators (QuickSim vs. AccuSim, in this case) will gives different levels of accuracy (logic vs. voltage) for different simulation time (fast vs. slow). The students are also required to simulate the standard cell design with AccuSim to document the performance benefit of the custom layout. The students then follow the tutorial to produce the layout from place and route software. This illustrates the difference in layout area versus the time and effort involved in producing the layout. At this point in the semester, having invested a great deal of time in the custom layout, the students well appreciate the benefits and costs of the computer-produced layout.

3.2 GRADING AND SCHEDULE

The design project is uniquely organized to mimic the market operation of the semiconductor industry: time-to-market is critical in that virtually all profit is made when a part is first introduced and the price is high. As soon as a competitor introduces its product, the price drops and profits are reduced. To emphasize this point in the VLSI project, points for single designs are awarded based on project completion, with more points being given for designs that are completed earlier. Projects are collected and graded every Monday. If the project is complete and correct at the first due date, the score is 11 points out of a possible 10. Each week delay results in 1 less point for the project, with 8 being the minimum score. Thus, students who keep up with the project and produce the correct circuit are rewarded with a higher score, much as companies that introduce products early are rewarded with higher profits.

3.3 DESIGN NOTEBOOK

To document his/her work, each student is required to maintain a design notebook, each page numbered and dated. As computer work seems to take an incredible amount of time, particularly if the student is not well prepared, I require specific information for each phase of the design process to encourage the student to be as efficient as possible when at the computer. For the schematic, the student must have a sketch of the schematic in the notebook before starting the schematic entry. For the simulation, the student must have the input and output waveforms sketched, including the time of breakpoints. For the layout, the student must have a stick diagram of the layout. Thus, when the student goes to the computer, he or she is ready to enter the correct data, completing the project as efficiently as possible.

SUMMARY
This paper describes a market-driven design project in an introductory VLSI design class. The project is uniquely modelled on the semiconductor industry, rewarding students for early completion of the project while learning the fundamentals.

REFERENCE

[1] S. Kang and Y. Leblebici, *CMOS Digital Integrated Circuits: Analysis & Design*, McGraw-Hill, Oct. 2002.

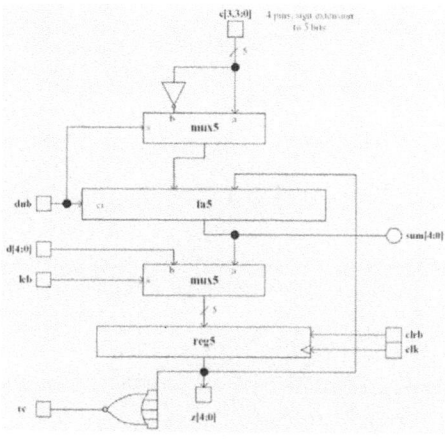

Fig. 1: Up/down counter schematic

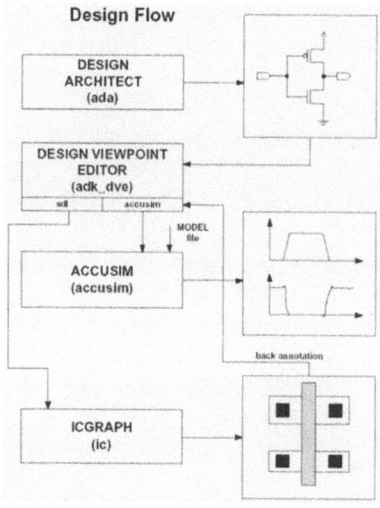

Fig. 2: Design flow using Mentor Graphics computer-aided design software

CONCURRENT SYSTEMS' HARDWARE DESIGN USING PETRI NETS

GOMES L., COSTA A.
lugo,akc@uninova.pt
Universidade Nova de Lisboa / UNINOVA
Faculdade de Ciências e Tecnologia
Department of Electrical Engineering
2829-516 Caparica
PORTUGAL

1. Introduction

In the past, when systems were mostly modeled through Finite State Machines (and Finite State Machines with data-paths modeled by Register Transfer Languages), the key concepts were global states and transitions between those global states. Although, for systems where concurrency is a key aspect, state machines proved to be inadequate, as they need to produce a model resulting from the combination of the state machines associated with the different sub-systems, leading to the well-known problem of state explosion and to the problem of adequate interpretation of the resulting states.

Petri nets [1] can be viewed as a generalization of a state machine, where several states can be active simultaneously, and transitions can start at a set of "states" and end in another set of "states". In this sense, Petri nets evolution is evaluated locality (and not globally as in state machines). Petri nets are directed graphs with two distinct sets of nodes: places, drawn as circles or ellipses; and transitions, drawn as bars, rectangles or squares. Places can represent states, but also resources. More generically, places are the passive entities, as opposed to transitions that model active entities. Petri nets do not give special emphasis to states, as state diagrams, nor to actions, as data flows: states and actions are both "first-class citizens".

Their graphical expressiveness, complemented by their formal verification capabilities, have been key features recommending their usage for applications where concurrency plays a key role.

Key formalisms to be used for system design in the future, where system-on-a-chip concept and billions of transistors in the same chip are major technology highlights, needs to support formal verification (as far as simulation is not enough for very complex systems) and extended support for concurrency modeling, namely synchronization mechanisms among concurrent processes, resource sharing modeling, critical sections, and so. Petri nets are, naturally, one candidate to play the role of the key formalism.

So, hardware design clearly benefit from the use of formal methods, namely from the use of Petri nets as modeling formalism. These should also provide a simple way to support system modifications and evolution.

A.M. Ionescu et al. (eds.), Microelectronics Education, 23–27.

Anyway, as far as digital hardware is concerned, we need to come to implementation. In this sense, we need to consider several issues like implementation specification, implementation language and implementation technology.

Again, Petri nets can be a valuable formalism as far as the same system's model that was used for system modeling and propriety verification, can now be used as specification for the implementation. Afterwards, this specification can be translated into a hardware description language, like VHDL [2],[3],[4], and synthesized in hardware. At this level, programmable logic devices, like Complex Programmable Logical Devices (CPLDs) and Field Programmable Gate Arrays (FPGAs), play a very important role, not only in the industry world, but also at the laboratory level, giving to the student the necessary flexibility to allow experimentation.

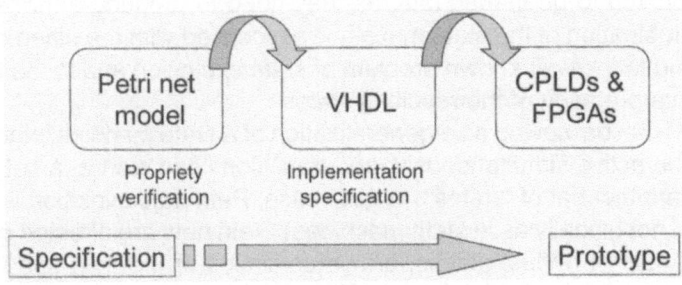

Fig. 1: From specification to prototype

The students of the Electrical and Computer Engineering Degree at Universidade Nova de Lisboa, Portugal, experience the presented roadmap (expressed in Figure 1) within one of their disciplines, named Digital Systems Design [5]. A similar experience (slightly shorter) is applicable for the students of the Computer Science Degree in a "sister" discipline [6].

2. Project syllabus

To illustrate the referred procedure flow, from specification to prototyping, we select a "family" of mini-projects, where we can have concurrent activity of several entities in the system. Also, if needed, is easy to built-up a series of mini-projects with specific characteristics, in order to allow each group of students to have a "unique" system to implement. The family of mini-projects is about parking lots controllers' implementation, where we can have several cars coming inside and leaving at the same time.

For each parking lot, we consider several entrances and several exits, one or several floors, and different combinations of paths between the different floors.

Figure 2 presents possible layouts for the parking lots. Also parking places at each level can be easily changed. To get inside or to leave the park is necessary to go through a gate, presenting a card or getting permission from the supervisor to proceed. Also, to circulate between levels we have sensors installed in the pavement in order to determine how many cars are present in each level, and issue recommendations about parking places availability through a set of output traffic lights.

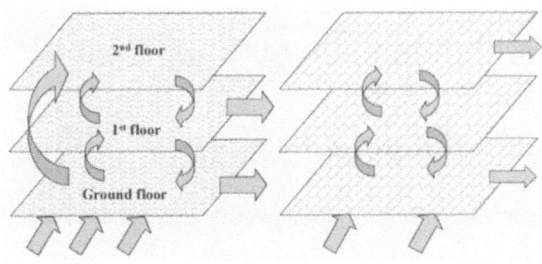

Fig. 2: Alternative layouts for our parking lots

3.   Modeling issues

The modeling process usually starts considering a parking lot with several entrances and several exits. Figure 3 presents a simplified model for a parking lot with 444 parking places. The model scales well, if necessary, to add new entrances or exits (as common with Petri nets). Afterwards, further refinements are carried out in order to accommodate the car movement between floors. In Figure 4, explicitly modeling of free and occupied parking places at first and second floors is complemented by a sub-model able to detect the movement of a car from the first floor to the second one (as an example), which is accomplished through the analysis of the evolution of sensor signals A and B (presence detectors). Students can model common situations (associated with well-behaved citizens) and also to foreseen recovery from anomalous situations, like go in the reverse way. Also, students can exercise and compare different sub-nets for modeling those characteristics and conclude about associated net proprieties (liveness, boundedness, safeness, complementary places, and so).

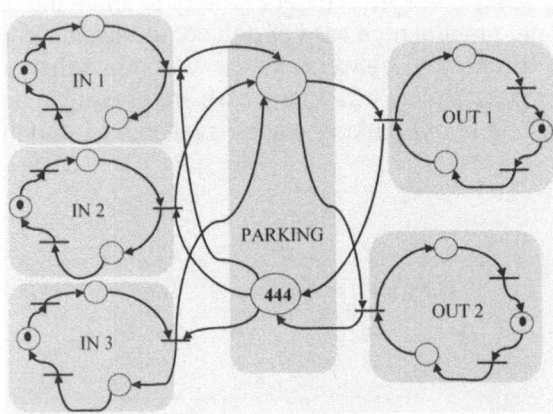

Fig. 3: Starting model for parking with three entrances and two exits

4.   Implementation issues and laboratory support

After modeling the system and perform its formal verification, the students go into VHDL coding and implementation using Xilinx Spartan-II FPGA kits, where the issues of correct implementation of conflicts are especially stressed.
The laboratory is composed by a set of workstations, connected to the web and to the discipline web site (containing all supporting materials and tools), equipped with workbench and adequate development environments (from Petri nets tools to Xilinx ISE environment for VHDL simulation and downloading into the kits).

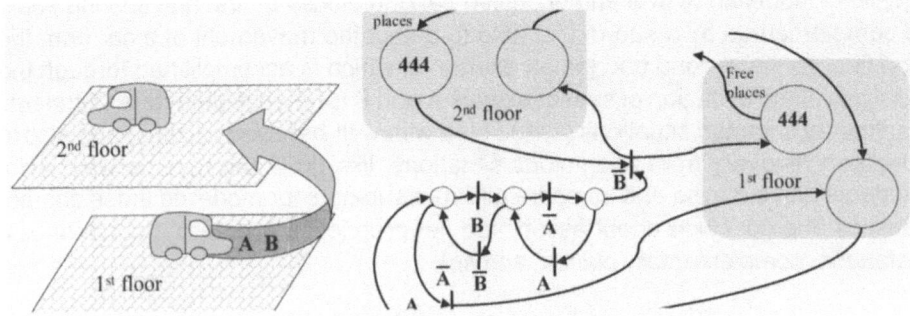

Fig. 4: Sub-model associated with car transition between two floors

Conclusions
Most of the students enjoyed the proposed project. It proved to be a good project to reach the initial goals: from specification to prototype, through formal verifica-

tion. Several students expanded the initial specification at their own expenses (extra hours work).

References

[1] Wolfgang Reisig; Petri nets: an introduction. Springer-Verlag New York, Inc., 1985.

[2] IEEE Std 1076-2000 (Incorporates IEEE Std 1076-1993 and IEEE Std 1076a-2000) IEEE Standard VHDL Language Reference Manual; 2000; ISBN 0-7381-1949-0

[3] Stephen Sjoholm, Lennart Lindh; "VHDL for Designers"; Prentice Hall; 1997; ISBN 0-13-473414-9

[4] Douglas L. Perry; "VHDL"; McGraw Hill Text; 1998; ISBN: 0070494363

[5] http://www-ssdp.dee.unl.fct.pt/leec/csd/20032004 – course on "Concepção de Sistemas Digitais" (Digital System Design), Computer and Electrical Engineering degree; Universidade Nova de Lisboa, Portugal

[6] http://www-ssdp.dee.unl.fct.pt/lei/sasd/20032004 – course on "Síntese e Análise de Sistemas Digitais" (Digital System Design), Computer Science degree; Universidade Nova de Lisboa, Portugal

FPGAS IN MICROSYSTEMS EDUCATION

POLLITT-SMITH H. AND XU S.
Canadian Microelectronics Corporation
210A Carruthers Hall, Queen's University, Kingston Ontario, K7L 3N6, Canada

ABSTRACT

Field programmable gate array (FPGA) technology has become an increasingly useful tool employed in universities for both teaching and research. In this paper, we present a series of FPGA-based platforms and CAD tool environments which have been supplied to the member universities of the Canadian System-on-Chip Research Network (SOCRN) by Canadian Microelectronics Corporation (CMC) to support microsystems education, targeting the specific areas of digital signal processing (DSP), embedded system and multimedia design. The various features of these platforms, plus design examples that may be used in a classroom environment are presented and discussed.

1.   INTRODUCTION

Programmable logic (specifically Field-Programmable Gate Arrays – FPGAs) is becoming an increasingly attractive research and teaching tool in universities, particularly for System-on-Chip (SOC) research and training. FPGAs are becoming more and more feature-rich, providing complete system design platforms with memory blocks, microprocessors (soft and hard macros), multipliers and DSP hard macros, embedded system (e.g., bus architectures and peripheral components) and application-specific (e.g., DSP and telecom) intellectual property (IP) blocks. To ensure that Canadian university educators take full advantage of this enabling technology, CMC, manager of the SOCRN – an advanced microchip research and design network that links over 30 Canadian universities across the country-- supplied 253 FPGA-based platforms to 28 member universities of the SOCRN. These systems, called System-Level Prototyping Stations (SLPS), are complete design and prototyping stations, including host PCs, a pre-installed CAD tool environment, and FPGA development boards with features that enable research and training in DSP, embedded system design, hardware/software co-design and co-verification, IP block design and verification, and multimedia application development. The reprogrammable nature of FPGAs makes these platforms desirable in an educational setting, as they can be re-programmed and reused across multiple projects and courses, resulting in maximum benefit to the university. The varied and application-specific interfaces on the boards, and system design environment (a "system to FPGA" flow) expand the SLPS capabilities beyond traditional VLSI (HDL-based) training to application-based system-level design.

A.M. Ionescu et al. (eds.), Microelectronics Education, 29–32.
© 2004 *Kluwer Academic Publishers.*

2.   SLPS for DSP

The first of the 3 platforms that CMC has distirbuted to universities is the SLPS for DSP (based around the Altera DSP Development Kit, Professional Stratix Edition) that features a Stratix EP1S80 FPGA, and high-performance analog-to-digital (A/D) and digital-to-analog (D/A) converters. This environment supports system-level design with Altera's DSP Builder, a Simulink blockset of DSP components (e.g., FIR filter, FFT, Reed-Solomon) that target implementation on Altera FPGAs. This blockset allows students to design and simulate a DSP system or block in the Matlab/Simulink environment, then implement their design on the FPGA (without having to hand-code DSP algorithms in HDL); real-time results can be viewed and analyzed in the Simulink environment in real-time or after the application runs on the FPGA board. Linking a ubiquitous system design environment (Matlab) directly to FPGA implementation expands the usability of FPGAs beyond the traditional VLSI research community. This platform can be used in a classroom to teach DSP system design in the Matlab/Simulink environment without requiring in-depth HDL experience.

For example, Figure 1 is a block diagram of a digital filter. It uses three numerically controlled oscillators (*NCO*) to generate sinusoidal signals at 1-MHz, 30-MHz and 50-MHz respectively. The signals are added together before passing through the digital filter. To study the effect of filtering, the *Filter* block can be configured as different types of FIR or IIR, low-pass, band-pass or high-pass filters. There are several *signalTap* analysis blocks inserted to capture signal activities on-board, and these signals are displayed in Simulink while the system runs at speed in hardware; it is not simulation but real-world signal analysis. The NCO and Filter blocks are programmed using NCO and FIR/IIR compilers that work with the Altera DSP Builder and Quartus II tools for FPGA implementation.

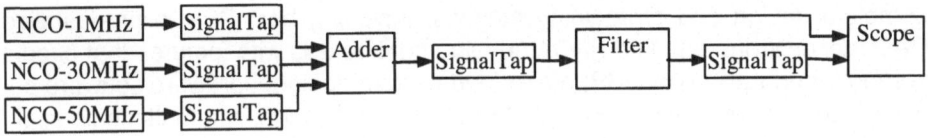

Fig. 1: Block diagram of a digital filter design in Matlab

In addition to the NCO and FIR/IIR compilers mentioned above, there are other reusable IP cores available with this board, including Reed Solomon, Viterbi and FFT compilers. With the large number of logic elements available on the FPGA and a library of DSP-related IP cores, students in a project course can design and validate complex DSP systems (for example, software defined radio) with a pair of these boards to implement transmit and receive functionality, or study hardware acceleration in a multithreaded, multiprocessor environment by designing a

DSP coprocessor for the system. As well, the DSP board has two independent banks of external SRAM memory that would support design involving concurrent memory architecture, and a daughter card connector for Texas Instruments' DSP boards to support projects in architectural exploration using different DSP processors.

3.   SLPS for Multimedia

The second platform that universities have access to through CMC is the SLPS for Multimedia (based around the Xilinx Multimedia Board), featuring a Virtex-II 2000 FPGA, and a variety multimedia interfaces such as video I/O, SVGA, 10/100 Ethernet, and audio. The various video interfaces make this platform attractive for image processing courses and projects, demonstrating, for example, the real-time operation of MPEG-4. The audio and Ethernet interfaces allow study of Voice over IP (VoIP) techniques. This system includes Xilinx's Embedded Design Kit (EDK) with the MicroBlaze embedded processor block and peripheral IP components.

Figure 2 is block diagram of a simple system design example built in EDK. The system contains a MicroBlaze (MB) connected to BRAM over the Local Memory Bus (LMB). The BRAM is a hard block on the Virtex-II. A UARTLite, external SRAM memory interface, and GPIO are connected to the On-chip Peripheral Bus (OPB). The General Purpose I/O (GPIO) can be used to connect to the LEDs or DIP switches on the board. MB and OPB peripheral blocks are available in the EDK design library. Users can add their own peripheral on the OPB bus, such as an Ethernet MAC (EMAC) (shown in Figure 2). If this example is used for a hardware design course, the EMAC can be a user-designed block with OPB interface. Students may use it to learn how to integrate a user-designed IP block (in HDL) into a system and verify it in the system environment. If this example is used for an embedded software design course, the EMAC also can be directly generated from EDK and students can focus on writing device drivers and application code, such as a web server application. Also, this example can be used in a HW/SW co-design course; students (perhaps in teams) may develop both HW and SW and get hands-on experience of co-design and co-verification in a system-on-programmable-chip environment. With the on-board Ethernet interface, the system can be implemented and both the EMAC design and application SW can be verified with real-time traffic.

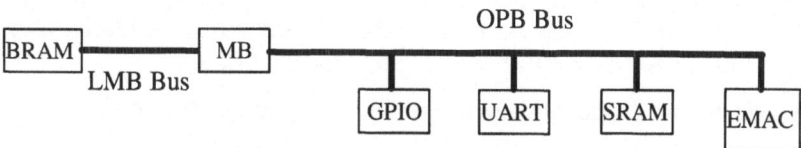

Fig. 2: Block diagram of a simple system-on-programmable chip design

As well, there are five independent banks of external SRAM on board, which can be used to support exploring memory concurrency to improve performance of multiprocessor architectures.

4. Embedded System Design Environments

The third platform that CMC supplied to universities is the SLPS for Embedded Systems, (based around the Altera Nios Development Kit, Stratix Professional Edition) featuring a Stratix IS40 FPGA, and a number of peripheral interfaces typical of embedded systems, such as 10/100 Ethernet, LCD, and serial port. This platform is useful for demonstrating concepts in computer architecture and could be used to study multithreaded multiprocessors; Altera's SOPC Builder toolset allows for the fast construction of an embedded system (including the Nios processor, and a library of peripheral IP components such as a UART and memory controller); instead of building a complete system from scratch, the hardware architecture, system library, and device drivers can be generated through the SOPC tools, then implemented and tested on the FPGA platform, using real-time data to exercise the system. This platform is also useful for designing and validating a reusable IP block, and for verifying that block's operation in a complete system. Rather than stopping at block-level simulation, students can prove their design by adding their block (e.g., Ethernet MAC) to the Nios system in the SOPC environment, and demonstrating its real-time operation.

SUMMARY

This paper presents three FPGA-based platforms and CAD tool environments that have been distributed to the member universities of the Canadian System-on-Chip Research Network (SOCRN) by Canadian Microelectronics Corporation (CMC) to support microsystems research and education. These environments are well suited in a classroom environment for teaching SOC concepts, HW/SW co-design and co-verification targeting digital signal processing, embedded systems and multimedia applications.

REFERENCES

[1] P. Lysaght, "Platform FPGAs", Winning the SoC Revolution: Experiences in Real Design, edited by G. Martin and H. Chang, Kluwer Academic Publishers, pp.141-158, 2003.

[2] J. kempa, et. al., "SOPC Builder: Performance by Design", Winning the SoC Revolution: Experiences in Real Design, edited by G. Martin and H. Chang, Kluwer Academic Publishers, pp.159-185, 2003.

[3] www.altera.com/products/devkits/altera/kit-dsp_stratix_pro.html

[4] www.altera.com /products/devkits/altera/kit-nios_1S40.html

[5] http://support.xilinx.com

THE PLATFORM AS AN INTERFACE IN A SOC DESIGN CURRICULUM

SANDER I., JANTSCH A., TENHUNEN H.,
Royal Institute of Technology, Stockholm, Sweden

1. TRENDS IN SYSTEM DESIGN

The steady increase of complexity of single chip systems drives the search for restricting the design space in meaningful ways. The trick is to restrict the design space such that the design process becomes fast and efficient while the resulting product is still close to optimal. One widely used way to do this is to provide architectural templates that allow to quickly assemble a large number of components in a very systematic way. The recent trend to platform based design [1,2,3] is emerging because it is increasingly difficult to organize a large number of pre-designed intellectual property (IP) blocks on chip. Nexperia [4,5] is a successful example of a platform. As figure 1 shows, a typical instance of the platform may consist of a RISC and a VLIW media processor, four buses, a central memory controller and a fairly large number of dedicated functional blocks. It is much easier to start with a platform like this and optimise and fine tune it for a

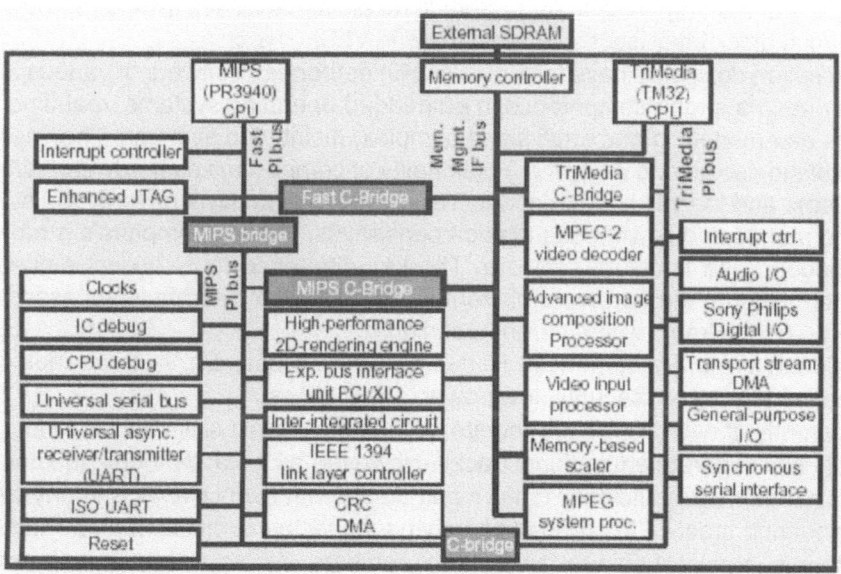

Fig. 1: **PNX8500 Viper processor based on the Philips Nexperia platform**

A.M. Ionescu et al. (eds.), Microelectronics Education, 33–37.
© 2004 *Kluwer Academic Publishers.*

particular product, than to start from scratch. Also, analysis and design tools can be developed for the platform and reused for every product development, which is a significant advantage because the development of specific tools is typically far beyond the possibilities of individual design projects.

2. CONSEQUENCES FOR EDUCATION

What are the skills needed when platform based design becomes the dominant style for designing SoC ASICs? Apparently we have to distinguish between *platform designers* and *application designers*.

Platform designers will resemble traditional hardware and ASIC designers and will need a strong competence in digital and analog hardware design, a solid knowledge of electrical and physical properties of transistors and wires, and a fair understanding of applications and the relevant system properties such as performance, cost and power. They will need an appreciation of how the platform is used by application designers and they will typically cooperate with application designers because platforms are usually developed with a particular application design project as driver. However, the key difference to the earlier VLSI generation is that the emphaise is on system level electrical issues instead of transistor or circuit block level issues as used to be in VLSI.

Application designers have to deal a lot with embedded software of various kinds. They need a strong competence in embedded operating systems, real-time systems and modelling and analysis of complex, distributed systems. They need to be able to design and verify the functionality of complex mixed hardware/software systems and the will have to conduct sophisticated analysis to verify the system performance and power consumption constraints. Here the emphasis is more on embedded and firmware software. The key difference to traditional embedded systems is that the hardware platforms are more complex (multiprocessor) and even reconfigurable with different techniques.

Obviously there are some joint issues for both groups, e.g. such as fault tolerance, architectures, security, low power, and low cost.

Apparently, these two groups require sufficiently distinct skills and knowledge to justify two separate educational tracks. However, as it stands today, the concept of platform is not sufficiently mature and stable with many overlaps in day-to-day engineering practice to propose entirely unconnected curricula. If a few standard platforms, stable of several generations, emerge, the education of platform designers and application designers can be separated further much like we have specialized curricula for hardware design and for software engineering with the processor as dividing platform. However, since this is not yet a reality we propose a connected curriculum with two specialization tracks.

3.   THE CURRICULUM

The program is a 1½ years Master of Science program and attracts students with computer science, computer engineering or electrical engineering background. Figure 2 gives an overview of the program structure and lists the individual courses. The numbers denote EECS credit units. The program is a further development of the SoC Master program [6], which is currently operated at KTH. The new program will become active in the fall 2004.

3.1   BASE BLOCK

First the students pass a common block of courses which takes approximately half a year and starts with an introductory course to Embedded System, which also provides a survey of many important topics many of which are elaborated much more in other courses later on.
SoC Architecture is one of the central courses in the program. It discusses all important SoC processor and communication architectures and presents severa

Master Thesis (30)		
Application Design		Platform Design
Elective Courses (15)		Elective Courses (15)
SOC-Applications (7.5)		Embedded Software (7.5)
Anatomy of CAD-tools (7.5)		Anatomy of CAD-tools (7.5)
Fault-Tolerant Systems (6)		Fault-Tolerant Systems (6)
SOC Modeling (7.5)		Electronic System Packaging (7.5)
System Verification (7.5)		Low Power & Mixed Signal IC (7.5)
		Radio Electronics (7.5)
Compulsory Course		Compulsory Course
Embedded Software (7.5)		Digital Circuit Design (7.5)
Embedded Systems (7.5)	Common Courses.	Digital System Engineering (7.5)
SOC-Architecture (6)	ASIC Design (7.5).	Digital Hardware Design (9)

Fig. 2: FThe curriculum consists of a common basic block (37.5 ECTS-credits); two alternative specializations with one compulsory course (7.5 ECTS-credits) and several elective courses (15 ECTS-credits); and finally a master thesis (30 ECTS-credits).

important platforms. The Digital Systems Engineering and Digital Hardware Design courses convey the basics in hardware design and the important transistor and wire abstractions. The latter is a necessary foundation to allow for sensible analysis of performance and power at the system level later on. The ASIC course introduces high level hardware design, synthesis, verification, simulation and testing techniques and tools.

3.2   PLATFORM DESIGNER

In the platform designer track the Digital Circuit Design course is compulsory. It is a continuation of the Digital Systems Engineering course and elaborates further on the physical properties of transistors and wires. In two −3 elective courses the student can study topics such as radio electronics, low power, fault tolerance, CAD tools of embedded software.

3.3   APPLICATION DESIGNER

The only compulsory course in the application designer track is embedded software. It introduces the topics layered embedded software architectures, hardware drivers, communication services, resource management and real-time operating systems. This course is an innovation and does not yet exist at KTH and perhaps nowhere else in this particular form. In 2-3 elective courses the student can further deepen his/her knowledge in advanced verification and modelling techniques, fault tolerance, CAD tools and SoC applications.

Summary
At KTH we have developed the successfully operating SoC Master Program further to meet the expected requirements on SoC engineers in the next few years. Based on the assumption that platforms will play a central role in SoC design in the near future, we have designed a curriculum consisting of two tracks, one for platform designers and one for application designers.

REFERENCES

[1]   Alberto Sangiovanni Vincentelli, "Defining Platform-based Design", EEDesign of EETimes, February 2002.

[2]   Henry Chang, Larry Cooke, Merrill Hunt, Grant Martin, Andrew McNelly, and Lee Todd, Surviving the SOC Revolution - A Guide to Platform-Based Design, Kluwer Academic Publishers, 1999.

[3]   edited by Grant Martin and Henry Chang, Winning the SoC Revolution, Kluwer Academic Publisher, 2003.

[4] J. Augusto De Oliveira and Hans van Antwerpen, "The Philips Nexperia Digital Video Platform", Winning the SoC Revolution, pp. 67-96, Kluwer Academic Publisher, edited by Grant Martin and Henry Chang, 2003.

[5] Philips Semiconductors. Nexperia. http://www.semiconductors.philips.com/ products/nexperia/

[6] L. Hellberg, A. Hemani, J. Isoaho, A. Jantsch, M. Mokhtari, and H. Tenhunen, "System Oriented VLSI Curriculum at KTH", Proceedings of the International Conference on Microelectronic Systems Educations, MSE97, 1997.

A NEW SYSTEM C-BASED FOUNDATION FOR THE CE CURRICULUM

SHANKAR R., JAYADEVAPPA S.
Florida Atlantic University, Dept. of CS&E,
777, West Glades Road, Boca Raton, FL – 33432

Abstract

A typical graduate of computer engineering (CE) program pursues a career in the computer industry or with a company that integrates computers into complex products. The Bachelor's degree curriculum in CE needs to focus in the future on system aspects and the integration of the hardware with software. The current curriculum introduces the hardware concepts with courses on processors, computer architecture, VLSI, electronics, and design automation [1]. Similarly, software concepts are addressed with courses on data structures and algorithms, operating systems, and software engineering. Though these courses met the earlier needs of the industry, we need to re-orient the courses based on the current and future industry requirements and job opportunities that are cross-disciplinary: A current embedded system warrants a seamless integration of software and hardware into a system that meets ever expanding functional and quality metrics. Under this notion, a system is more than the sum of its parts, that is, software and hardware. This requires a holistic approach and a constant dialog between software and hardware practitioners.

1.   Introduction

Computer Engineering is considered to be the confluence of electrical engineering (EE) and computer science (CS) . In fact many of the courses are taught by EE and CS faculty members, with effort made to bring in the concepts of relevance to CE. A separate identity is needed for CE today, thanks to the exponential growth in the complexity of the embedded systems [2] and a consumer market that is expanding rapidly. Today's embedded systems, as exemplified by cell phones, PDAs, and videogames, may have multiple processors, tightly integrated software and hardware, and real-time power efficient operating systems. This is neither the field of EE nor CS majors, but a distinctly different type of graduate, viz., a CE major, with enough depth in both hardware and software in different dimensions. These dimensions are: computing and communication, design and verification, abstraction and implementation, and finally, user and engineering requirements.

A fresh computer engineering graduate would find jobs that will place him or her at the software-hardware interface, viz., development of device drivers, prototyping, protocol development, and system testing and verification. Architectural evaluation, real-time performance and power management, still owned by the prior generation EE majors, will have to accommodate the new generation of CE

A.M. Ionescu et al. (eds.), Microelectronics Education, 39–44.

majors and their new tool suite, as we elaborate below. Application development may continue to involve CS majors, while algorithmic development and mixed signal hardware design may involve EE majors.

2.   Proposed CE Curriculum

Concurrency and complexity of the embedded systems demand a new paradigm for the CE curriculum [see 3, 4]. We believe that an integrated software-hardware environment for modeling, analysis, design, verification, emulation, and implementation is needed for the students to understand the system-wide impact of an isolated decision they might make or implement in one area.

Figure 1 presents our proposed integrated system design, verification, and development environment [5], which can form the basis for a new CE curriculum [see 6 for an example]. The five columns identify the five major elements in the engineering of an embedded system. However, their placement next to each other also implies the possibility of integrating them to address system/sub-system considerations. We use this model to identify the contents of different courses and their role in computer engineering curriculum based on an integrated environment that we have developed. One can develop a set of courses addressing each of these five issues. Horizontal strips could be used to identify system level courses, such as software hardware co-design, performance evaluation, unified verification flow, and mixed level embedded system design.

The topics that need further elaboration, relative to the currently existing curriculum, would be software-hardware co-design, embedded system design, real-time systems, RTOS, system design flow, concurrency modeling, performance evaluation, networks on chips, verification, design for tolerance, and design productivity enhancement. We base these suggestions on our industrial experience and feedback, as well as the technology roadmap [2].

We have developed an integrated environment with the ability to support software hardware co-design: SHINE (Software Hardware Integrated Design Environment) [5]. We developed this to enhance system design productivity. We have since then realized that this would also be an excellent vehicle to integrate the various concepts and courses that a CE student needs to get exposed to, just as VLSI and Software Engineering have been proposed to form the basis, respectively, for EE and CS majors. SHINE is based on SystemC, a system level design language [7]. SystemC is based on the C++ language and has constructs to support hardware modeling. The language supports multiple levels of abstraction, a common environment for design and verification, and hardware-software co-design. Currently the SystemC language is undergoing standardization, but has already been adopted by over one hundred design companies. The infrastructure requirement is low as SystemC is open source. A standard ANSI C++ compiler is sufficient to develop and simulate models in SystemC.

Figure 1, Column (a) addresses the software part of the co-design flow. It involves the development of application software which will be executed on the hardware architecture developed. A 'Data Structures and Algorithms' course will form the basis for software related design. Traditional software courses need to be enhanced as follows: Operating System: Include real-time operating system; and Software Engineering: Include concurrency and real-time issues. Note that concurrency programming with C++ is rapidly evolving, thanks to the Internet Client-Server concurrency [8,9]. Since SystemC is a concurrent version of the C++ language, one can take advantage of the concepts developed in such books. New courses on device driver development and application software development will be needed. We have not seen any good books on that. SystemC/SHINE provides operating system primitives. Thus, modeling and analysis of the real-time effects of various scheduling, inter-process communication and context switching functions would be feasible computer-science oriented projects.

Figure 1, Column (b) depicts the verification techniques needed at various levels of system modeling abstraction [10]. SystemC is a good choice as it supports verification through the SystemC verification extension [7]. Assertions based verification will be supported in SystemC in the future with additional constructs. Two good books that would help in this domain, perhaps developed because of the Internet needs, are listed at the end [11,12].

Figure 1, Column (c) displays   the hardware part of the co-design. Course on modeling using an ISS (instruction set simulator) will enhance the ability of the students to comprehend software testing in the appropriate embedded system environment, before the actual hardware is ready. In SHINE an ISS is integrated to support the development of general purpose software applications.

 Figure 1, Column (d) presents the ability to support communication among the different architectural components present in a system. Communication among these components can be modeled at various levels of abstraction depending on the level of detailing necessary. In the future, the communication sub-system may contain both local synchronous buses and global asynchronous links [2]. This is very similar to the Internet links. Thus, the courses on Internet communication and protocol stacks could be adapted for this.

Figure 1, Column (e) identifies issues pertinent to the modeling and design of various peripheral devices present in a typical system. This will include components such as, GPT (general purpose timer) and SPI (serial peripheral interface). A new course here would stress   IP (intellectual property) development, for ease of integration and reuse [1].

The broken line boundaries in the vertical strip show current technology gaps. To develop a seamless flow one needs to develop suitable converters, integrators, translators, and/ or wrappers.

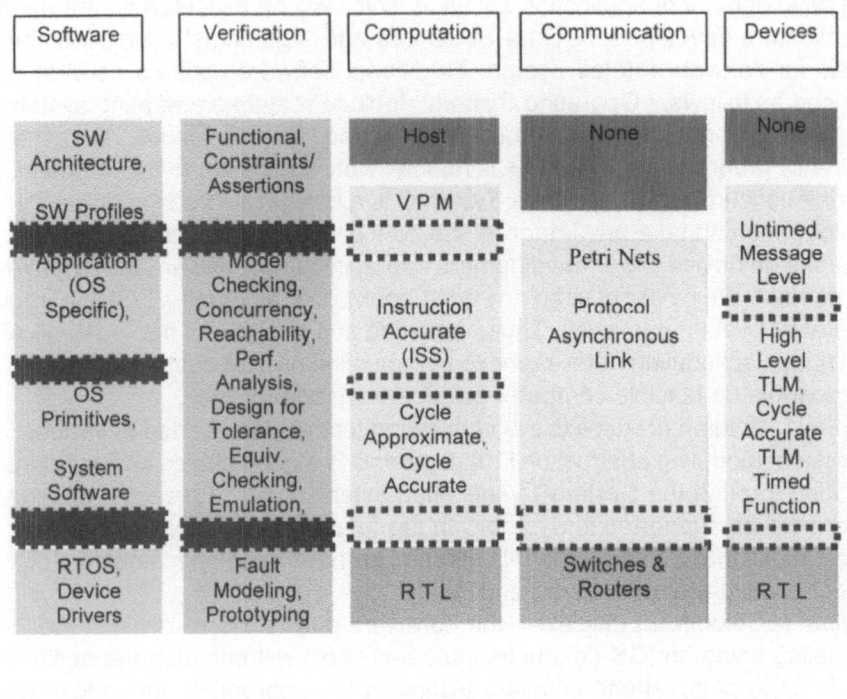

(a)   (b)   (c)   (d)   (e)

Fig. 1: Various Components of Software Hardware Co-Design and Verification
Legend : TLM – Transaction Level Modeling, VPM – Virtual Processor Modeling, OS – Operating System, SW – Software, RTOS – Real Time Operating System, RTL – Register Transfer Level.

Figure 2 presents the different computer engineering courses that will address the industry and hardware software co-design and co-verification requirements. SHINE methodology being developed at our University will provide the necessary environment for software hardware co-design using components at various levels of abstraction. A course on System Level Modeling should concentrate on modeling at higher levels of abstraction above that of RTL (Register Transfer Level). The course on software hardware co-design will enable the students to learn the hardware software interaction early in the design cycle. Projects can be developed to explore architectural variations.

Conclusions

This paper presents different aspects of the current CE curriculum that need to undergo a change to bridge better the gap between the background of the graduating engineer and the needs of this decade's embedded systems industry. The earliest this new cadre of graduates will arrive at the industry gates is four years away. We propose that the open source SystemC and our SHINE infrastructure for software-hardware co-design provide an easy way to develop the foundation on which all or most of the CE curriculum can be based.

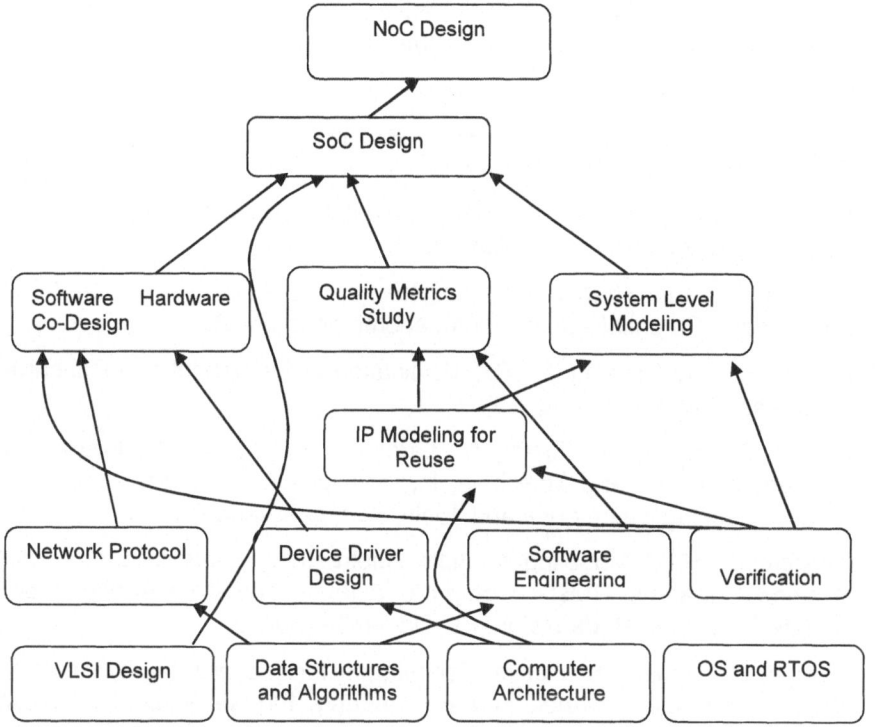

Fig. 2: Courses for the Computer Engineering Curriculum
Legend : NoC – Network on Chip, SoC – System on Chip, VLSI – Very Large Scale Integration, IP – Intellectual Property

References:

[1]   Bouldin,D., ``Enhancing System-level Education with Reusable Designs '' Proceedings of European Workshop on Microelectronics Education (EWME), Kluwer Academic Publishers, pp. 5-8, Aix-en-Provence, France, May 18, 2000.

[2]   www.public.itrs.net

[3]   Jantsch, A., Modeling Embedded Systems and SOCs: Concurrency and Time in Models of Computation, Morgan Kaufmann Publishers, 2004.

[4]   Jantsch, A., and Tenhunen, H., Networks on Chip, Kluwer Academic Publishers, 2003.

[5] Jayadevappa, S., SHINE: An Integrated Environment for Software Hardware Co-Design, Ph.D Dissertation, Computer Science and Engineering, Florida Atlantic University, Boca Raton, FL, December 2003.

[6] Curriculum, BS in Computer Engineering, Florida Atlantic University, Boca Raton, FL, www.cse.fau.edu/undergraduate.htm

[7] www.systemc.org

[8] Hughes, C., and Hughes, T., Parallel and Distributed Programming Using C++, Pearson Education, Inc., 2004.

[9] Bacon, J., and Harris, T., Operating Systems : Concurrent and Distributed Software Design, Addison Wesley, 2003.

[10] Freytag, G., and Shankar, R., "Digital Hardware Verification Methods : Existing and Potential Applications", submitted to EWME '04.

[11] Apt, K., R., and Olderog, E., R., Verification of Sequential and Concurrent Programs, Springer_Verlag, 1997.

[12] Berard, B., Bidoit, M., Finkel, A., Laroussinie, F., Petit, A., Petrucci, L., Schnoebelen, Ph., and Mckenzie, P., Systems and Software Verification Model-Checking Techniques and Tools, Springer_Verlag, 1999.

[13] Jillellamudi, H., " Modeling Multiple Abstraction Levels in SoC's Using SystemC", Mater's Thesis, Computer Science and Engineering, Florida Atlantic University, Boca Raton, FL, December 2003.

[14] Karnati, R., "Survey of Design Techniques for Signal Integrity", Master's Thesis, Computer Science and Engineering, Florida Atlantic University, Boca Raton, FL, December 2003.

POSTER PRESENTATIONS

TEACHING CUSTOM AND AUTOMATED CELL DESIGN

BOULDIN D-W.,TAN C., PATEL K-J.
Electrical & Computer Engineering, University of Tennessee, Knoxville, TN
37996-2100 dbouldin@tennessee.edu

1.   Introduction

Custom design and verification of leaf-cells for standard-height and bit-slice librar-ies can be used to provide students with experience in performing these physical level tasks manually. We describe a one-semester graduate course [1] in which students compare their own custom layouts with automated results in terms of ar-ea, delay and design time. The availability of valid automated solutions provides the students with targets that serve as feasible bounds on the layouts and also inspire competition between the student and the design automation software. Projects are combined into TinyChips and submitted for fabrication via MOSIS [2] for testing during the subsequent semester.

2.   Design and Verification Flow

The Cadence Design Systems [3] custom integrated circuit design bundle was selected to support the laboratory assignments in this course because it provides students with an integrated flow for custom design and verification. As shown in part of Figure 1, the designer begins by entering a hierarchical schematic to spec-ify the desired logic and the W/L of each transistor. The resulting net-list is then simulated using detailed transistor models (Spectre or SPICE).

A.M. Ionescu et al. (eds.), Microelectronics Education, 47–52.
© 2004 *Kluwer Academic Publishers.*

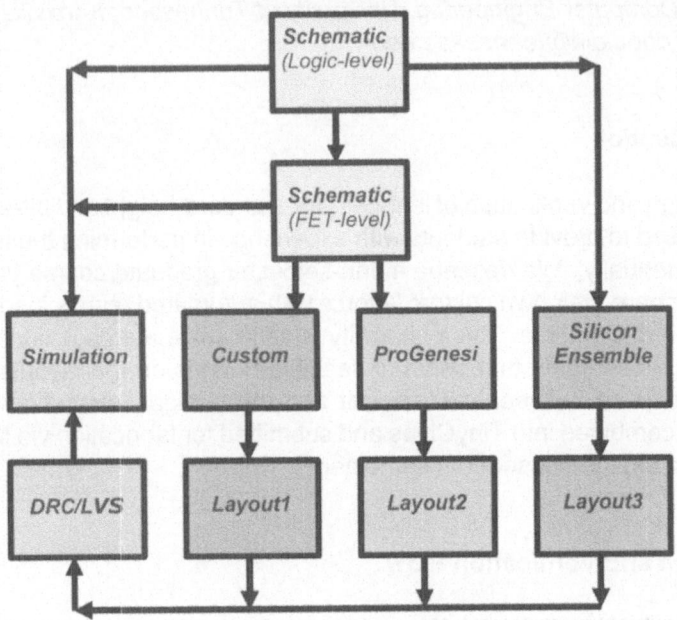

Fig. 1: Design Flow

For manual or custom implementations, a layout editor (Virtuoso) is used to construct the transistors with the proper W/L, place them and route them to correspond to the schematic net-list. The layouts are performed to conform to the
desired geometry constraints (e.g. standard-height, bit-slice, analog, etc.). A design rule check is then made followed by an extraction to trace the final net-list
and to calculate the interconnect resistances and capacitances. The final net-list
is compared to the original schematic to ensure they correspond. Post-layout simulations are then performed to determine the final timing delays assuming a standard load with four inverters in parallel.

The metal layers and pins of the standard-height leaf-cells developed by the students are abstracted and then appended to an existing standard-cell library so
that automatic placement and routing software (Silicon Ensemble) can incorpo-
rate these additional cells into a design. This step ensures that the manual layouts
adhere to the desired height so that power and ground rails will exist when cells
are placed adjacent (actually overlapping slightly). This step also ensures that the
input and output pins within each leaf-cell can be accessed correctly for routing
additional interconnect layers automatically (e.g. grid spacing of 8-lambda hori-
zontally for metal-2 and 10-lambda vertically for metal-3 in the AMI 0.5-micron
CMOS process). A manually drawn standard-height cell is shown in Figure 2a.

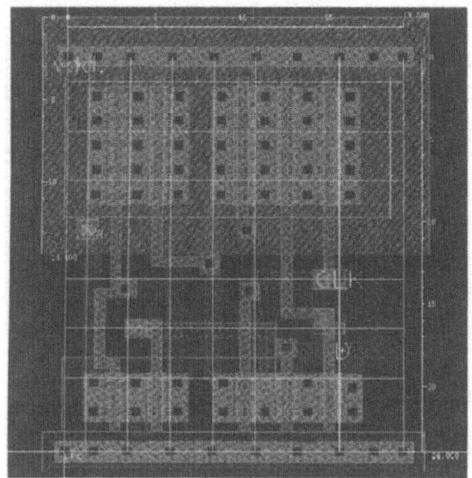 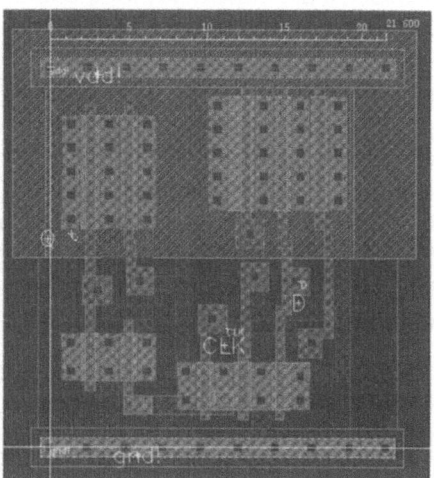

Fig. 2: . Standard-height cell layouts: (a) Manual and (b) Automatic

ProGenesis [4] is an automatic leaf-cell generator. As shown in Figure 1, ProGenesis accepts the same transistor-level net-list as Virtuoso (after a format conversion). The tool also reads a process rule file so that it is cognizant of the desired spacings and thus produces layout that is free of geometric design rule errors. It can also read in a constraint file so that the layout can be compliant with standard-height constraints for power and ground spacing as well as pin locations. The resulting layout is then verified in the same manner as a custom layout. For bit-slice leaf-cells, a color plan must be followed so that power, ground and control signals will automatically form buses when the cells are tiled in a vertical dimension to form a data-path with the desired number of bits in a data-path. Signals enter each cell from the left and exit to the right.

3.   Projects

Typical projects for this course include counters, shift-registers, flip-flops, multipliers and dividers. These digital circuits are suitable for both standard-height and bit-slice formats. The automated layouts using ProGenesis are performed first to serve as targets for the students when manually composing their standard-height leaf-cells. These targets provide hints to the students but also inspire competition between the student and the design automation software.

Table 1 below shows a comparison of various projects in terms of area in square microns and delay in nanoseconds. The last column in the table shows the percent difference between the automatic and manual layouts. This value was calculated by taking the product of area and delay for the automated layout minus the product of area and delay for the manual layout divided by the area-delay product for the manual case. The results show that the manual and automated layouts differ by 15 percent or less and that ProGenesis appears to be better for more complex circuits. Since the students are novices at both using ProGenesis and in producing manual layouts, these results should not be used to make definitive judgments about the quality of the tool. However, the availability of the automated solutions is a powerful aid to learning (and grading). The students readily observe that the automated layouts are generated within minutes (sometimes taking up to two hours on a Sun Enterprise 220) with almost no effort on the part of the student whereas manual layout of circuits like these generally requires at least ten hours of tedious labor.

Circuit Number	Complexity (Trans.+Nets)	Manual Area	Auto Area	Manual Delay	Auto Delay	Percent Difference
1	20	461	518	239	244	15
2	27	899	1037	6244	6232	15
3	41	1037	1181	209	184	0
4	57	1267	1315	861	884	7
5	65	1660	1830	2598	2584	10
6	67	1728	1505	405	462	-1
7	77	2419	2087	780	811	-10
8	84	1901	1613	194	202	-12
9	85	1728	1723	536	521	-3
10	86	2354	2117	413	404	-12
11	104	2242	2247	1036	1003	-3

Table 1. Comparison of Manual and Automatic Designs

4.   Comparison of Bit-Slice and Standard-Height Designs

In addition to comparing their custom standard-height leaf-cells with those produced automatically using ProGenesis, it is also desirable for each student to compare his/her bit-slice data-path design to one produced automatically using standard-height placement and routing. Since we have yet to learn how to get ProGenesis to produce layouts conforming to the bit-slice constraints, each student is asked to tile his custom single bit-slice into an 8-bit data-path and measure the resulting area and delay using a standard load (fan-out of four inverters). He then uses his single-bit standard-height cell and Silicon Ensemble to produce an 8-bit macro consisting of several rows of standard cells. In some cases, the data-path layout has half the area-delay product of the standard-height macro layout. For example, one student's data-path had an area-delay product of 22 units (square microns * nanoseconds * 10 **5) while the corresponding standard-height macro was 53 units.

5.   Silicon Verification via MOSIS

For verification, student projects are submitted to MOSIS for fabrication. To facilitate testing, the students are provided with a Smartframe which consists of a pad-frame AND a built-in self-test (BIST) capability. Student projects can be placed inside the frame as shown in Figure 3 and stimulated using patterns generated on-the-fly at full system speed.

The responses are collected and compressed into signatures which are then compared on-chip to hardwired golden values obtained from simulation. Thus, the golden values are compared with the live signatures to determine if all is well. If not, raw inputs corresponding to the functional vectors for each student design can be presented to the external pins and the raw outputs observed for debugging purposes. Most of the student projects returned using the Smartframe have operated as expected. We attribute this high success rate to this methodology which encourages thorough simulation prior to fabrication.

The Smartframe and a 12-cell standard-height cell library are available at no charge from the University of Tennessee. Composer schematics can be used to interconnect multiple standard cells from this library which can then be automatically placed and routed using Cadence Silicon Ensemble. The resulting layout can also then be viewed using Cadence Virtuoso.

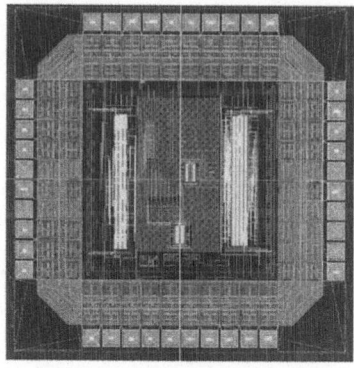

Fig. 3: Smartframe with four student layouts and CMP fill.

ACKNOWLEDGMENTS

We gratefully acknowledge Cadence Design Systems, Inc. and Prolific, Inc. for support of the laboratory with their integrated circuit design tools and MOSIS for fabrication of prototypes for our classes at no charge. We also thank the students of the 2002 section of ECE 651 for providing the design results of their projects.

REFERENCES

[1]   ECE 651, University of Tennessee,
       http://vlsi1.engr.utk.edu/ece/bouldin_courses/

[2]   MOSIS, http://www.mosis.org/

[3]   Cadence University Program, Cadence Design Systems, http://
       www.cadence.com/company/university/

[4]   Prolific, Inc., http://www.prolificinc.com/

[5]   Smartframe and Standard-Cell Library, University of Tennessee,
       http://vlsi1.engr.utk.edu/ece/cadence.html

A DSP/FPGA/FPAA BASED APPROACH FOR ELECTRONIC EMBEDDED INSTRUMENTATION

CHIABERGE M., CARABELLI S., ROLANDO P., DAMILANO C., FERRARESE L., GARZELLA R., GOBETTO L., CHIAPUSSO L., FILIPPONE F.
Politecnico di Torino, Interdepartmental Mechatronics Laboratory,
Corso Duca degli Abruzzi 24, 10129 TORINO, ITALY
Tel. +39 0115646239, Fax. +39-0115647963, email: marcello.chiaberge@poli-to.it

1.   INTRODUCTION

AKU (Actuator Kontrol Unit) is the prototype of a new embedded computing platform, based on state of the art electronic devices (DSP+FPGA+FPAA).
AKU Based Instrumentation is a simple and modular architecture for fast prototyping and development of electronic embedded test equipments.

2.   PLATFORM ARCHITECTURE

AKU platform is made up of a HOST and TARGET architecture.
AKU HOST framework consists of a SERVER and CLIENTS. The SERVER concentrates the CLIENT applications requests and forwards them to the TARGET.
AKU TARGET framework is composed of 3 elements: software (real-time software and user tasks on DSP), firmware (FPGA) and hardware (FPAA and Field Modules).
Figure1 shows the block diagram of the AKU Platform

.

A.M. Ionescu et al. (eds.), Microelectronics Education, 53–58.
© 2004 *Kluwer Academic Publishers.*

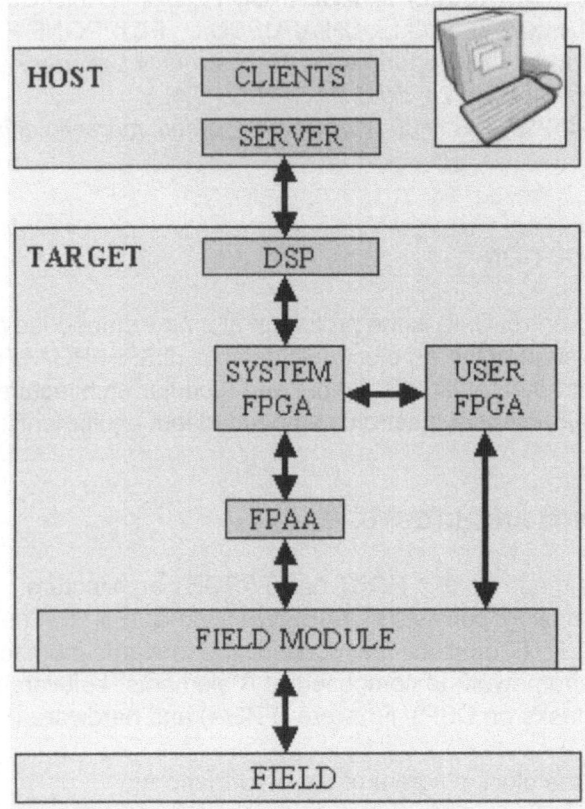

Fig. 1: AKU Platform

2.1 HOST

The HOST allows the user to configure and manage AKU Instrumentation through a virtual interface running on different platforms such as PC and PDA. AKU can be also managed through remote interfaces (WLAN, TCP/IP, etc…).
Several tools, grouped in the "RTC Suite", were developed to show and monitor signals in real-time, tune parameters and manage DSP/FPGA/FPAA parameters/ status.
The SERVER allows the HOST to communicate to the DSP and to download the real-time software/firmware through a serial communication.

2.2 AKU

The AKU consists of two electronic boards: DSP board and FPGA/FPAA board. The DSP is the supervisor of the entire system; it monitors signals from field, generates commands properly and communicates with the HOST.

The FPGA operates as peripheral I/O extension of DSP functionalities, as well as coprocessor. The user establishes its behaviour instantiating Intellectual Properties (IPs), which interact to the field (addressing digital I/Os) and/or improve DSP computing capability.

A particular locked section of the firmware architecture manages the communication between DSP and IPs in memory-mapped mode.

The FPAA satisfies the request of configurable analog components for real time programmable analog conditioning using switched capacitance technology.

2.3   FIELD MODULE

The Field Module is a special purpose interface board; it serves as hardware interface between the AKU and the field, performing the required actions and conditioning signals to send back to AKU. For instance, the features of a Field Module can be optical isolation, uncoupling functions, signal conditioning, voltage and current control, etc.

The AKU communicates with Field Modules through digital and analog signals, grouped in the AKU Field Bus, which can be customized by the user through a simple and flexible assignment.

The open structure and such numerous available I/Os of AKU Field Bus allow plugging on the different Field Modules so that it is satisfied the demand of quite different applications.

3.   MODULAR ARCHITECTURE

AKU Instrumentation is based on a modular architecture. Changing the application field, the user can modify the DSP software application, the user firmware and the Field Module custom hardware, whereas HOST tools, RTOS and the AKU remain the same for all kind of applications. Figure2 shows a Field Module device plugged into the AKU.

Embedded Instrumentation and modular architecture are key elements in developing today's customer applications, which often require a reconfigurable and flexible hardware and portability over different platforms.

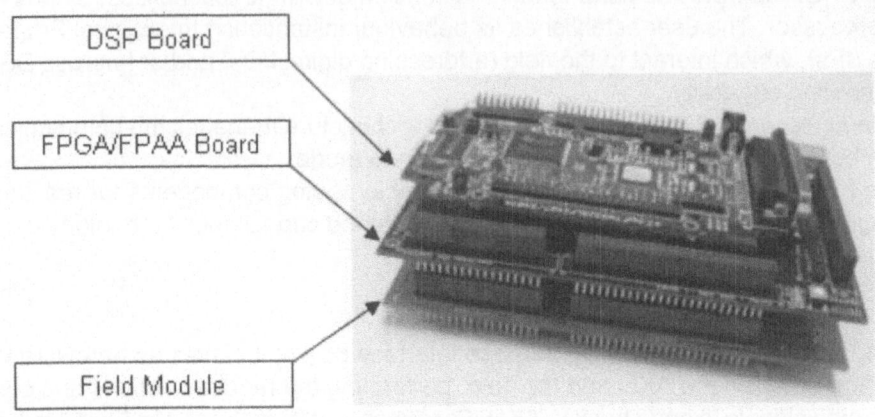

Fig. 2: A picture of the AKU-system

4.   APPLICATION FIELD EXAMPLES

AKU is ideal for several high-performance applications, especially for rapid proto-typing of industrial controls and monitoring platforms.

It can also be used to realize embedded, portable, low power and low cost field-instrumentation, that can reach the accuracy of laboratory instruments and maintain flexibility (number of available options and parameters) through programmability.

AKU concept could be surely a precious resource for class labs.

A basic level lab can be targeted to practice singularly FPGA, FPAA and DSP, gain valuable experience on different state of the art electronic devices and learn the fundamental principles of programming (hardware, firmware and software).

At more advanced levels, AKU could be considered as an integrated system (FPGA+DSP), which students can handle to analyze the communication strategies.

Finally, complex test rigs can be prepared adding Field Modules and connecting plant and sensors: asynchronous and brushless motors (current, speed and position control loop), telecommunication measurement and testing, power characterization are the most significant examples of applications.

The extreme flexibility of the AKU can also be useful to achieve easy and rapid prototyping in academic research; the integration of programmable logics /analogs, DSP and hardware resources would stimulate interdisciplinary collaborations.

5.   PROJECT IN PROGRESS

The two following projects are been developed on AKU and at present are under test:

- BERT: a Bit Error Rate Tester that evaluates the $P(e)$ on transmission lines (optical fiber, coaxial, etc.). A high performance programmable logic soldered on BERT Field Module allows a high throughput, 8 different bit-rates (from 125Kbps up to 1.25Gbps), 5 different PRBS lengths, Clock and Data Recovery for asynchronous receiving, standard I/O differential LVPECL DC. The AKU configures the BERT functions accordingly with parameters chosen by user and processes the data results (error bit number and time of measurement) to show it to user.

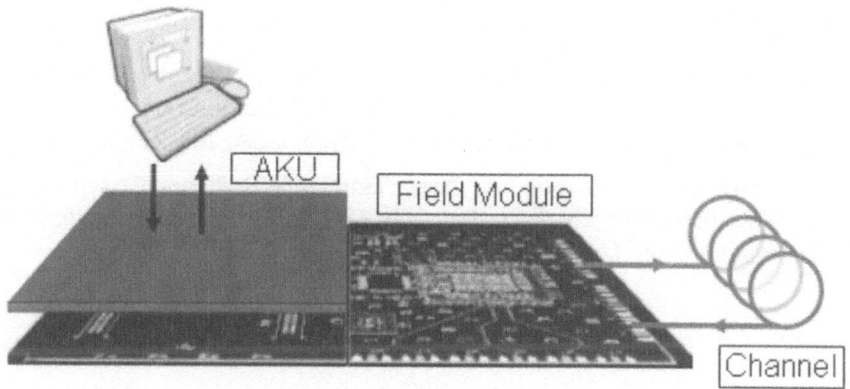

Fig. 3: BERT framework

- IPS: the Intelligent Power Supply system is intended to provide a configurable and controlled power profile to devices under test (power characterization of batteries, supplies, drives). A simple interface allows the user to configure different operating modes and their parameters, such as voltage and current ranges, power sourcing or sinking. The AKU serves both as control and monitoring unit, reading the instantaneous operating status from IPS Field Module and logging acquired data and real time calculations results.

Fig. 4: AKU instrumentation for power application

REFERENCES

[1] M. Chiaberge, W. Santero, D. Amerio, "Digital Solutions for reProgrammability", SCI'2001, The Fifth Multi-Conference on Systemics, Cybernetics and Informatics, Orlando, Florida, USA, July 22 - 25, 2001.

[2] B. Jeong, S. Yoo, S. Lee, K. Choi, "Hardware-Software Cosynthesis for Run-time Incrementally Reconfigurable FPGAs", Asia South Pacific Design Automation Conf. (ASPDAC), pp.169-174, Jan. 2000.

DIGITAL HARDWARE VERIFICATION METHODS

Existing and Potential Applications
FREYTAG G., SHANKAR R.
Florida Atlantic University, Dept. of Computer Science and Engineering,
777 Glades Road, Boca Raton, FL 33431, USA

ABSTRACT
The ultimate goal of system-on-chip (SoC) verification is to obtain the highest possible level of confidence in the correctness of a design. But the complexity of SoCs is growing exponentially, even as the market is pressuring design cycle times to decrease. The dual challenges of increasing complexity and decreasing time are creating an urgent need for the application of advanced verification methods. With this in mind, we present a technique for identifying both existing and unexplored applications of digital hardware verification methods. The technique is patterned after the periodic table of chemical elements and is applicable to verification of other types of hardware and software as well. Our periodic table indicates which methods are the most broadly applicable and hence the most worthwhile to learn. The table also suggests which applications have potential for research.

1.   INTRODUCTION

SoC verification has many aspects, and different tools or methods are typically used to address each aspect. The various aspects of SoC verification can be represented by verification metrics, which indicate whether or not a device meets its specified requirements. The applicability of each method can be expressed in terms of the metrics that the method is capable of measuring. Our periodic table of digital hardware verification is aimed at revealing the range of applicability for each method and the ways in which different methods can double-check or even substitute for each other.

1.1   THE PERIODIC TABLE CONCEPT

The inventor of the periodic table of elements noted that certain groups of elements shared similar chemical properties. This observation led him to arrange the known elements in a table such that elements with similar properties appeared together in columns. Each row in the table represented elements having the same total number of electron shells. There were empty spaces in the table when it was first introduced, because at that time there were no known elements with certain combinations of chemical properties and electron shells. However, the table theorized the existence of these elements, and they were eventually discovered.

59

A.M. Ionescu et al. (eds.), Microelectronics Education, 59–64.

1.2 A PERIODIC TABLE FOR DIGITAL HARDWARE VERIFICATION

Our periodic table (see Figure 1) lists the commonly used methods of digital hardware verification in rows and the metrics in columns. The methods are arranged according to the level of design detail at which they are most often applied, from high-level at the top to low-level at the bottom. The metrics are shown in the order in which they are typically measured during the design cycle, from early stages on the left to late stages on the right. Each combination of method and metric is assigned one of three ratings:

- Useful – either the method is commonly used in industry to obtain the metric, or the method has at least been demonstrated to be useful in measuring it;
- Not useful – the method is impractical for measuring the metric extensively because of cost or time constraints, or the method does not analyze the design at a level where the metric is measurable;
- Unexplored – the method does not appear to be commonly used for obtaining the metric, but it has significant potential for doing so.

	Metrics early stage → late stage	Output	Reachability	Assertions	Performance	Latency	Code coverage	Power	Logical equivalence	Fault coverage	Sensitivity/ tolerance
high-level	Software-based simulation	−	−	−	−	−	−	−	—	—	−
	Model checking	—	−	−	?	?	−	?	—	—	—
	Equivalence checking	—	−	—	?	?	−	?	−	—	—
	Hardware-based acceleration	−	−	−	−	−	—	?	—	?	—
	Emulation	−	−	−	−	−	—	?	—	?	—
	Prototyping	−	—	—	−	—	—	—	—	—	—
	Static timing	—	—	—	—	—	—	—	—	—	—
	Fault grading	—	—	—	—	—	—	—	—	—	—
low-level	Signal integrity	—	—	—	—	—	—	—	—	—	—

Legend: − = useful, — = not useful, ? = unexplored

Fig. 1: Periodic Table for Digital Hardware Verification

The ratings reveal both the existing and potential applications of each method. In particular, the "Unexplored" ratings indicate aspects of verification where non-traditional methods could be applied. Such applications are analogous to the undiscovered elements that existed at the time the original periodic table was created.

2. VERIFICATION METHODS AND METRICS

In the sections below, we analyze the applications of each method to the metrics shown in the table.

2.1 SOFTWARE-BASED SIMULATION

Software-based simulation is the most versatile of all verification methods. Theoretically, any system or process that can be represented in software can be verified in simulation. However, when a large design is modeled in software, simulation becomes less practical as the degree of detail captured in the model increases. Software-based simulation is therefore most useful in obtaining early-stage metrics.

2.2 MODEL CHECKING

When compared with software-based simulation, model checking has the advantage of capturing the structure and functionality of a design in a mathematical (rather than executable) form. The design can then be analyzed without the need to generate and run many individual tests. Model checkers are already used to verify reachability and assertions, but a mathematical model of a system should also be able to predict performance, latency and power consumption. These predictions could then be compared with the results of software-based simulations as a sanity check.

2.3 EQUIVALENCE CHECKING

Equivalence checking is typically a point solution used after synthesis of RTL code to verify that the synthesized netlist is logically equivalent to the code. Like model checkers, equivalence checkers build mathematical models to represent designs. However, equivalence checking requires that two models be built, one for the RTL code and one for the netlist, so that they can be compared for logical equivalency. Perhaps two such models could also be compared to verify equivalent performance, latency and power consumption. This would help confirm that the synthesis was optimal.

2.4 HARDWARE-BASED ACCELERATION

Hardware-based acceleration has much of the versatility of software-based simulation along with much higher speed, but it runs simulations at the gate level.

Hardware accelerators are therefore potential tools for obtaining any metric that is meaningful at the gate level. Such metrics could include estimates of fault coverage and power consumption, based on signal-level activity.

2.5   EMULATION AND PROTOTYPING

Emulation and prototyping are similar in principle to hardware acceleration and could provide some of the same metrics. However, prototypes tend to be built from one or more FPGA devices, which generally do not afford much visibility of the internal signal-level activity of the design.

2.6   STATIC TIMING ANALYSIS

Synthesizers use static timing analysis to optimize logic paths at the gate level. Timing analyzers calculate the longest propagation delay between any two registers in the synthesized netlist. This calculated delay provides a more accurate indication of the device's performance and latency at the signal level than any other method considered here. Since timing analysis must be done in any case when creating an optimized netlist, there is no extra effort involved in obtaining performance and latency figures at this level.

2.7   FAULT GRADING AND SIGNAL INTEGRITY ANALYSIS

Fault grading and signal integrity analysis are usually accomplished with point tools late in the design cycle. However, as mentioned previously, hardware accelerators may also be useful in estimating fault coverage. Preliminary sensitivity and tolerance data can be obtained from software-based simulations and finalized as the chip layout is created.

DISCUSSION
Existing research and industrial expertise suggests ways to maximize the usefulness of the methods described above. Peterson [4] has derived an equation to estimate the efficiency of a verification method when applied to a design of a given size for a specified number of test cycles. His findings support the use of software-based simulators in the early stages of development, when the scope of testing tends to be limited. As the design is integrated and testing intensifies, hardware-based acceleration becomes the method of choice. Finally, when software is run on top of the hardware model, requiring even more test cycles, emulation is the preferred method.

However, when different methods are applied at different stages of testing, it becomes all the more important to ensure a smooth transition between stages. Murray [3] describes a synthesizable testbench that creates a bridge between software-based simulation and emulation. At first, the testbench creates test cases for system integration in the simulator. Once these tests are working properly, both the design and the testbench can be ported to the emulator. On the emulator, the testbench helps to automate and accelerate system-level testing. When a test fails on the emulator, it can be debugged in the simulator where the design is fully observable. The synthesizable testbench therefore allows the simulator and emulator to enhance each other's efficiency.

Girczyc [2] notes that assertions can also help bridge the gaps between different verification methods. Software-based simulators can monitor assertions at every clock cycle of every simulation to ensure thorough testing. Model checkers can verify the properties of a design by using assertions as constraints on inputs and as targets for outputs. Assertion metrics also indicate the level of coverage achieved in the verification process.

The efficiency of verification can be further increased through the use of different methods on different parts of a design simultaneously. Bailey [1] cites the case of an SoC developer who divided a design into three components: processor models, running on a high-level simulator; peripheral blocks, mapped into an emulator; and system memory, modeled in an RTL simulator. Compared to the option of using an emulator with the capacity to model the entire design, the developer achieved simulation speeds 500 times faster while reducing the verification cost by a factor of 20. Optimal application of a method thus depends not only on the metrics being measured, but also on the characteristics of the design architecture.

CONCLUSIONS

The periodic table concept has helped us identify the most versatile verification methods for digital hardware, along with a number of unexplored applications of specific methods. Both academic and industrial experience would seem to suggest that software-based simulation, model checking, hardware-based acceleration and emulation offer the broadest range of application of the methods considered here. These four methods are therefore not only the most worthwhile to learn about, but also the most potentially fruitful for research. In the future, we hope to study the unexplored applications in more detail to help create more efficient verification processes for digital hardware.

REFERENCES

[1] Brian Bailey, "Co-Verification: From Tool to Methodology," white paper, www.mentor.com, June 2002.

[2] Emil Girczyc, "Assertion-Based Verification Streamlines Design Outsourcing," EEdesign, Oct. 25, 2002. www.eedesign.com/story/OEG20021025S0034.

[3] David Murray, "Synthesizable Verification IP Speeds Design Cycle," EEdesign, Mar. 31, 2003. www.eedesign.com/story/OEG20030331S0061.

[4]   Gregory D. Peterson, "Predicting the Performance of SoC Verification
       Technologies," Proceedings of the VHDL International Users Forum Fall
       Workshop, Oct. 18-20, 2000, pp.17-24.

FROM MSI MODULES TO MICROPROCESSORS: FILLING THE GAP WITH PROGRAMMABLE LOGIC DEVICES

GOMES L., MALO P., COSTA A.
lugo, pmm, akc@uninova.pt
Universidade Nova de Lisboa / UNINOVA
Faculdade de Ciências e Tecnologia
Department of Electrical Engineering
2829-516 Caparica
PORTUGAL

1.   Introduction

Introductory digital system design within Computer and Electrical Engineering Courses is normally covered by one to three disciplines in the course curricula, during the first year or during the first two years.

In general terms, we may identify different phases within the teaching process.

On one hand, the first well-established step normally includes Boolean algebra and related issues and ends-up with the MSI-module (Medium Scale Integration integrated circuit module) view of the digital system. At this stage, it is common to emphasize the structural view of the system and associated schematics representation. The commonly used building blocks include from simple gates up-to common modules like adders, comparators, multiplexers, flip-flops, and counters. Behavioral descriptions of the system are introduced either for combinatorial logic using truth tables and algebraic expressions, or to sequential circuits using state diagrams. Common development tools include schematics editor, state diagram editor, logic simulators.

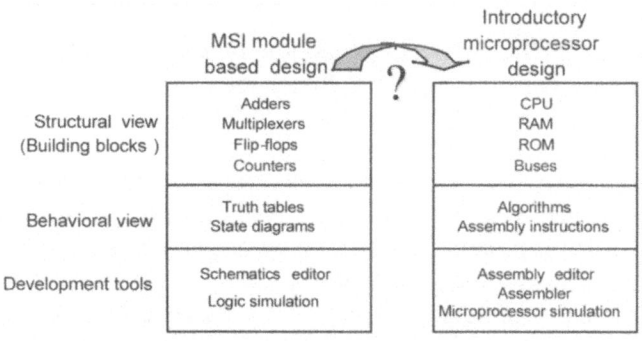

Fig. 1: Module-based and microprocessor-based frameworks

A.M. Ionescu et al. (eds.), Microelectronics Education, 65–69.

On the other hand, we can find the introductory microprocessor design phase. At this stage, it is common to emphasize the behavioral description of the system, accomplished through an algorithm, and complemented by the associated coding using assembly instructions. The algorithm is described using a specific notation for algorithm representation (from flowcharts, to pseudo-code, including some other graphical structured representations). Structural views of the system are used as a support for describing the system functionality, and include complex blocks, like CPUs, RAM, and ROM, interconnected using specific buses.

Between these two well-established phases, in some curricula, we can find a gap, as far as the student never faces a sound explanation of the microprocessor-based systems' functionality in terms of the MSI-module-based systems characteristics. So, as shown in Figure 1, one question remains: how to fill the gap between these two phases? At Universidade Nova de Lisboa, we fill this gap building a transition ([1]) from the "MSI-module" phase to the "microprocessor" phase based on the exploitation of the system's functional decomposition into data part and control part. We start at the MSI-module level, using common tools for this level, and at the end, before entering into the "regular" microprocessor-based design, the student has a project integrating the different views and tools of both environments, resulting in the merging of the columns presented in Figure 1. In this sense, we start with a MSI-module's vocabulary, concepts and tools and end-up with microprocessor terminology, using the whole set of tools, namely from schematics and logic simulators, to text editor, assembler and microprocessor simulators. In the next section, we present the plan for laboratory assignments supporting the referred bridge between the two "islands".

2.   Laboratory assignments plan

We chose to present the sequence of laboratory assignments (instead of details on lecture contents, which are close to these ones, of course), as far as it gives a concise view of our pedagogical approach to fill the gap between MSI-module based design and microprocessor design.

As referred, system's decomposition into control and data parts is a central concept in our approach. We start with a simple system project and end-up with a small microprocessor project. The laboratory assignments sequence stands for half a semester, and is composed by the following mini-project sequence:

- Introductory project: the goal is to implement a two-byte multiplier through successive additions. It is clearly a project close to the MSI-module student's background that introduces Arithmetic and Logic Unit (ALU) usage within register transfer architecture. Supporting development environment is introduced. Lab classes stands for two weeks (two classes of two hours each per week).
- Intermediary project: the goal is to implement a simple calculator, able to compute an expression like 2.A-B. With this project we introduce the

ALU reconfigurability capability that is fundamental for microprocessor operation. Lab classes stands for one week.

- Preparatory project: the goal is to get acquainted with memory usage, namely write/read data to/from RAM and read data from ROM (fundamental blocks within microprocessor architecture). Lab classes stands for one week.

- Mini-microprocessor project: the goal is to build a didactic mini-microprocessor, named 7_ [3]. Data part block decomposition of 7_ is presented in Figure 2. Lab classes stands for two weeks and a half. Students need to dedicate some extra-hours to conclude the assignment.

It is clear that for solution effectiveness it is important to have as much as possible a uniform development environment for the sequence of lab assignments. We consider that it is useless to argue in favor of Programmable Logic Devices (PLDs) usage, as it is almost mandatory to have PLDs to support the presented sequence of mini-projects. Those arguments (mainly implicit arguments, due to space constraints) led us to select a CPLD as the core of implementation platform and Xilinx ISE as the development environment. Schematics are used as the main form for hardware representation in these projects. In a latter point in the course curriculum, the same mini-microprocessor project will be used to support VHDL experiencing [2].

Considering that we don't want to build an entire set of tools for the new processor (assembler and simulator, at least), we decided to structure our 7_ as a sub-set of the widely used, well-known 8051/31 micro-controller family (not considering the "controller-oriented" parts of the device). In this sense, we have access to adequate assembler and simulators already in the public domain.

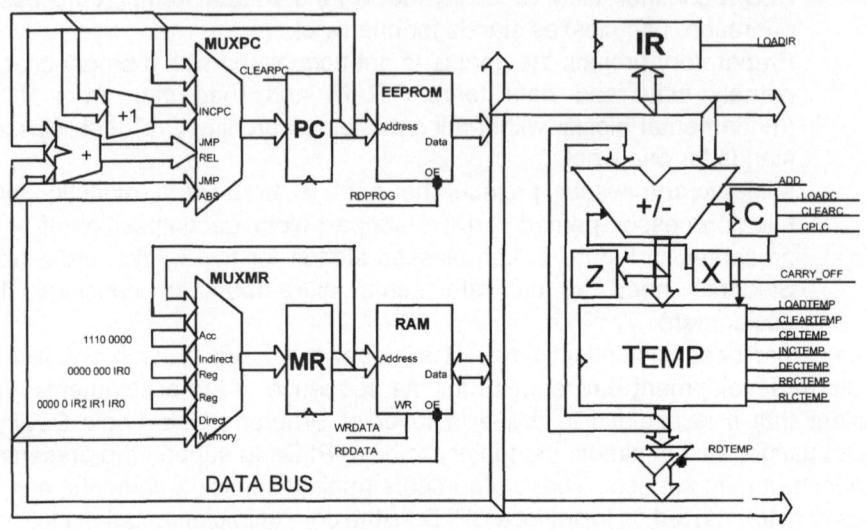

Fig. 2: Data part block decomposition of the didactic microprocessor 7_

3.   Laboratory support

In general terms, the laboratory assignments for digital systems disciplines at our Faculty (including those introductory disciplines) try to combine several empha- ses, namely:

- One emphases on the computer aided engineering component, using a set of engineering software tools (from specific editors to specific compilers).
- Other emphases on the workbench instrumentation component, to shorten distances and eliminate fears associated with the control of physical devices.
- Another emphases on system simulation, more and more important along the course (and real life) as the system complexity increases.

For that, the laboratory is structured to support working teams of three students. Students have access to ten work places and to a general-purpose programmer (for EEPROM configuration, for instance). A networked PC and an experimental set-up compose each work place. The PC has access to the Internet, including the discipline site, which contains the supporting materials (from datasheets to lab assignments descriptions and associated tools). The experimental set-up in- cludes power supply and testing resources (clock, inputs and outputs), and spe- cific breadboarded systems, ranging from a simple breadboard ready to receive some new experiment to some pre-prepared experiments, like the one presented in Figure 3 that supports the sequence of laboratory assignments referred in this paper. At the left hand side, we can identify two boards with general-purpose testing resources. At the right hand side, we can see the CPLD and JTAG programmer adaptor, EEPROM and RAM, specific for the proposed laboratory assignments. Every work place has one of these set-ups.

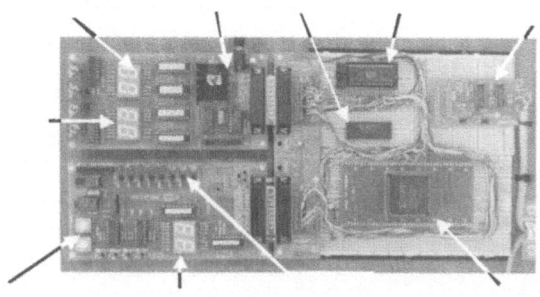

Fig. 3: Supporting experimental set-up

Conclusions
The paper presents a sequence of lab assignments (and implicit associated theoretical contents) that can be seen as a teaching module amenable to be used for filling the gap between two other common modules at introductory digital system design, namely the MSI-module based design and the microprocessor design. This module extensively uses programmable logic devices (namely CPLDs/FPGAs, RAM and ROM) and proved to be effective to get students involved and to bring them from their initial knowledge, mostly based on a structural view of MSI-module based architectures, to microprocessor design.

References

[1]   http://www-ssdp.dee.fct.unl.pt/leec/sl2 Digital Systems 2 web page

[2]   http://www-ssdp.dee.fct.unl.pt/leec/csd Digital Systems Design web page

[3]   "Introducing microprocessor design using programmable logic devices and public domain tools"; Luís Gomes; Proceedings of the 4th European Workshop on Microelectronics Education - EWME 2002; ISBN 84-267-1325-4

FORMAL SPECIFICATION FOR HARDWARE VERIFICATION

SHOJAI H., NAVABI Z.

Electrical and Computer Engineering Department, Faculty of Enginnering,
University of Tehran, Tehran, Iran, shojai@cad.ece.ut.ac.ir
*Nprtheastren University, Boston MA02115, navabi@ece.neu.edu

Abstract
In this paper we describe a methodology for the formal verification of a processor using the CTL property language. Processors are important in design and verification of digital systems because they have structures that represent most digital systems. Processors are programmable, have control parts, data parts and are rich in bus structure. Verification of CPU structures requires verification of data components, controllers, datapath and instruction level verification. This work uses a processor to discuss various features of formal verification. Because of generality of processors, we will be able to cover most aspects of property-based verification and properties used for this purpose.

1. Introduction

Formal verification [1], [2] is the process of checking whether a design satisfies certain requirements (properties). Processors are important in design and verification of digital systems because they have structures that represent most digital systems. Processors are programmable, have control parts, data parts and are rich in bus structure. Verification of CPU structures [3], [4], [5], [6], [7] requires verification of data components, controllers, datapath and instruction level verification.

This paper uses a processor to discuss various features of formal verification. Because of generality of processors, we will be able to cover most aspects of property-based verification and properties used for this purpose.

The simple CPU example discussed here is SAYEH (Simple Architecture, Yet Enough Hardware) that has a register file that is used for data processing instructions. The CPU has a 16-bit data bus and a 16-bit address bus. The processor has 8 and 16-bit instructions. Short instructions may contain shadow instructions, which effectively pack two such instructions into a 16-bit word.

The rest of this paper is organized as follows. Basic register verification is presented in section 2. Section 3 describes short memory and register file verification. Logic unit verification is presented in section 4 and section 5 and section 6 describe datapath and controller verification respectively. We conclude the paper in section 7.

A.M. Ionescu et al. (eds.), Microelectronics Education, 71–76.
© 2004 *Kluwer Academic Publishers.*

2.   Basic Register Verification

The *SAYEH* processor is a digital system with register and combinational parts. This Section covers verification of all vector (one dimensional) registers of *SAYEH*. We begin with the simple ones like the *Program Counter* and show verification of the *Instruction Register, Status Register and the Window Pointer*.

SAYEH Program Counter is a 16-bit register that is always clocked. Verification of this simple register is achieved by use of *CTL* [2] operator 'AG' and 'AX' as equation 1.

$$AG((in = n) => AX(out = n)) \qquad (1)$$

This says that the in value at time t0 becomes the out value at time t1.

SAYEH Instruction Register is a 16-bit register that is directly placed in *SAYEH* datapath. As shown in equation 2, verification of *IR* consists of an imply operator that conditions clocking of the register with the *IRload* input of the register.

$$AG((Irload \ \& \ in = n) => AX(out = n)) \qquad (2)$$

The *Status Register of SAYEH* is a two-bit register with *Carry* and Zero flags. We will use the same approach for verifying this register as that used for *IR*. Properties shown in equations 3, 4 verify loading of *Carry* and Zero flags. In these properties i can be '0' or '1'.

$$AG((SRload \ \& \ Cin = i) => AX(Cout = Cin)) \qquad (3)$$

$$AG((SRload \ \& \ Zin = i) => AX(Zout = Zin)) \qquad (4)$$

3.   Short Memory and Register File Verification

A component found in many RT level designs is a short memory or a register-file. Our example processor, SAYEH, is one such design. The data path of this processor includes a 64-word long register-file, the access to which is through a 4-word window. Verification of a register-file is important because its structure and application are unique and different from other RT-level components. In many ways, verification of a short memory or a register-file is similar to memory testing for static and dynamic memories. Right and left pointers to the register file are formed by adding the 6-bit base address to 2-bit offsets. Because of the size of these offsets (Laddr and Raddr), left and right pointers to the memory are never more than 3 locations apart. In order to verify this we first form the following diff macro to find the positive difference of it's a and b operands.

$$\#define\ diff\ (a,b) \{(a > b)?(a-b):(b-a)\}$$ (5)

The following property uses diff to check the maximum separation of these pointers.

$$AG(diff\ (Laddress, Raddress) <= 3)$$ (6)

A simple check for memory write is to make sure that the next state of a memory word that is being written into becomes the same as the memory input. The following property verifies memory writing.

$$AG((Laddress = i\ \&\ Lwrite\ \&\ in = n) => AX\ (memoryFile\ [i] = n)$$ (7)

It is important that when we write into a memory location, data is written only to the location that is being addressed and all other locations remain unchanged. The physical fault in large memories that causes a write into a location other than that being addressed is called coupling fault. We use a similar terminology for design errors that lead to addressing unwanted memory locations. Performing coupling check on our RegisterFile module means that writing into location i keeps all other locations, j, where   untouched. The property shown below verifies this. In this property i and j can be 0 to 63.

$$AG((Laddress = i\ \&\ Lwrite\ \&\ in = n\ \&\ \&\ MemoryFile\ [j] = m\ \&\ i != j)\ =>$$
$$AX\ (MemoryFile\ [i] = n\ \&\ MemoryFile\ [j] = m\ )$$ (8)

4.   Logic Unit Verification

The focus of this section is on verification of combinational circuits. The two major combinational parts of SAYEH processor are its ALU and address logic, which will be treated here. These parts are described at the behavioral level and their verification will only consider their primary ports, i.e., treating them as black boxes. This section shows how a complex behavioral description can be verified without having to get involved in its internal details.

SAYEH ALU is a 16-bit combinational logic. Its data inputs are 16-bit A and B and its data output is aluout. The ALU has control inputs that determine the function it performs. In addition, the cin input is its carry input used in arithmetic operations. Control outputs of this unit are zout and cout that are the zero and carry flags respectively. Properties for checking the basic operations of the ALU are shown below.

```
AG  A15to0 -> aluout = A);
AG( B15to0 -> aluout = B);
AG( Bleast -> aluout = B[7:0]);
AG( AandB -> aluout = (A & B));
AG( AorB  -> aluout = (A | B));
AG( notB  -> aluout = ~B);
AG ( AmulB -> aluout = (A[7:0] * B[7:0]));
```

Fig. 1: ALU Properties

5.   Datapath Verification

This Section looks at the components and their functions from a higher level of hierarchy. We verify micro operations of the datapath of *SAYEH* in which components discussed in the previous sections are instantiated. Micro operations in the datapath consist of placement of data into busses, transfer of data from one internal register to another, and routing of data read from the memory into data register and busses. In this section we verify some micro operations in *SAYEH*. Also we will show verification of some micro-operations that use the system memory. For this purpose the simple memory model will be used.

For the instruction fetch the Program Counter output drives the memory address bus and the data bus output of the memory becomes the input of the instruction register. Property shown below verifies this operation. The first use of the allzero macro insures that all Addressing unit control inputs are zero which causes the PC output to pass through to the Address bus. The second use of this macro guarantees that the Databus remains undriven by datapath busses.

```
//P10: Fetch
always ((allzero({ResetPC, PCplusI, PCplus1, R0plusI,
R0plus0}) &&
         allzero({Address_on_Databus, ALU_on_Databus}) &&
ReadMemory) &&
    (Databus == Memory(Addressbus, ReadMemory)) -> (IR.in
== AU.PC.out));
```

Fig. 2: Instruction Fetch Property

6.   Controller Verification

As compared with data components and the datapath of a digital design, the controller has a different style of coding and has different properties to verify. A typical controller has several states and it is important to be able to verify its flow and state transitions.

SAYEH controller is a state machine that issues control signals to the datapath components. There are five states in this machine that include reset, fetch, decode, exec and halt. Figure 3 shows the corresponding state diagram.
In this section we show properties for verification of SAYEH controller that is made of a *Finite State Machine (FSM)*.

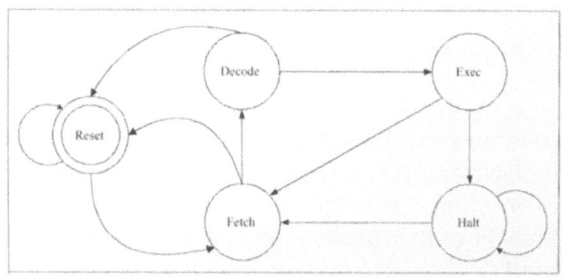

Fig. 3: Controller Block Diagram

DIRECT EDGES. Specific edges in the diagram of Fig. 3 can be verified by standard *CTL operators*. For example for verifying that there is an edge from state *RESET* to state FETCH, the following expression is used:

$$AG(RESET => EF(FETCH)) \tag{9}$$

DEADLOCK. Deadlock occurs if there are no conditions under which transitions are made out of a certain state. In this case, if the machine enters the deadlock state, it will stay in that state forever. As shown in the state diagram of Fig. 3, there are no dead lock states in the SAYEH controller. To verify this circuit that a state, such as *RESET*, is not a deadlock state, the following property should be used.

$$AG(RESET => EX(\sim RESET)) \tag{10}$$

REACHABILITY. A way of validating a state machine is to make sure that for any given state, there is at least one direct edge from another state into this state. In other words, all states are reachable from other states. These properties can be verified by standard *CTL* operators. For example for verifying that state *FETCH* is reachable from state *RESET*, the following expression is used:

$$AG(RESET => EF(FETCH)) \tag{11}$$

SPECIFIC EDGES. In addition to the general properties discussed above, there are other properties that can further verify the design of the state machine of Fig.3.

As shown in Fig.3, there is an edge from state *FETCH* to state *DECODE*. If we start in state *FETCH*, in the absence of *ExternalReset* the next state of the machine becomes *DECODE*. This is verified by the following property.

$$AG(FETCH \& \& \sim External\,\mathrm{Re}\,set) => AX(DECODE)) \qquad (12)$$

Conclusion

In this paper we covered some basics of developing properties for components of a design. Also we discussed verification of multi-level combinational circuits. We presented ways of verifying controllers and data paths.

The newest full version of this paper is at "http://cad.ece.ut.ac.ir/~shojai/paper/CPU_Verification.pdf"

References

[1]   E. M. Clarke, E. A. Emerson, and A. P. Sistla, "Automatic Verification of Finite-State Concurrent Systems Using Temporal Logic Specifications," in ACM Transitions on Programming Languages and Systems, 8(2), pp. 244–263, 1986.

[2]   Jae-Young Jang, In-Ho Moon, Gary Hachtel, "Iterative Abstraction-based CTL Model Checking.",in Design, Automation & Test in Europe, Paris, France, March 27-30, 2000.

[3]   Tom Schubert, "High Level Formal Verification of Next-Generation Microprocessors", 40th Conference on Design Automation Conference (DAC'2003), 2003.

[4]   J. R. Burch, "Techniques for Verifying Superscalar Microprocessors", Proceedings of the 33rd Design Automation Conference (DAC'96), 1996.

[5]   H. Choi, B. Yun, Y. Lee, and H. Roh, "Model Checking of S3C2400X Industrial Embedded SOC Product", Proceedings of the 38th Conference on Design Automation Conference (DAC'2001), 2001.

[6]   R. Jhala, and K. L. McMillan, "Micro Architecture Verification by Compositional Model Checking", Proceedings of the 13th International Conference on Computer Aided Verification (CAV'2001), 2001.

[7]   Goel, and W. R. Lee, "Formal Verification of an IBM CoreConnect Processor Local Bus Arbiter Core", Proceedings of the 37th Conference on Design Automation Conference (DAC'2000), 2000.

AN FPGA BOARD USED FOR DIGITAL LOGIC LABS

LEE S.
Department of Electronic and Electrical Engineering, Pohang University of Science and Technology, Pohang 790-784, South Korea.

1. INTRODUCTION

Digital logic design is an essential part of the curriculum of a university's electrical engineering department. With the increasing importance of digital devices in today's society and the increasing complexity of the digital devices used, most electrical engineering departments offer more than one course on digital logic design. The digital logic courses offered typically consist of an introductory digital logic course followed by one or more courses focusing on aspects of advanced digital logic design. Such advanced digital logic design courses involve the use of hardware description languages, field-programmable logic arrays (FPGAs), and electronic design automation (EDA) computer-aided design (CAD) tools, including back-end tools such Synopsis' logic synthesis tool [1], and front-end tools, such as Exsedia's Nimbus [2], used for graphical high-level design.

An essential part of an advanced digital logic design course is the lab, which should provide the students with the opportunity to practice designing complex digital logic circuits of various types using commercial software and hardware tools. In order to view the results of their design efforts and to experience the practical difficulties of actual hardware design, it is also essential for the students to be able to implement their circuits in hardware. In recent years, with increasing advances in FPGA technology, it is now feasible to use FPGAs to implement fairly large student projects such as 32-bit microprocessor designs.

A practical problem that arises when attempting to use FPGAs in a class lab is the availability of an FPGA board that can be used to prototype student designs. If sufficient money is available, a viable option is to purchase an FPGA prototyping board, such as the XESS board offered for use with Xilinx FPGAs [3]. However, such boards are costly and lag behind the state-of-the-art FPGA models offered. Also, if this type of "ready-to-use" FPGA prototyping solution is used in a university class, what is to distinguish the class from a short-term (e.g., one day to two weeks) industrial seminar? A university class should focus on the underlying theory, as well as the practical implementation aspects, involved with the use of tools and devices such as FPGAs. Thus, it would be more desirable if we could simply issue state-of-the-art FPGA chips to the students and have them create their own FPGA prototyping boards, using knowledge about how to connect to and configure the FPGAs based on information provided in data sheets.

A compromise solution is to design and fabricate a set of custom-designed printed circuit boards with state-of-the-art FPGA chips, and then use these FPGA boards to prototype student designs. Such a compromise solution may be

A.M. Ionescu et al. (eds.), Microelectronics Education, 77–80.
© 2004 *Kluwer Academic Publishers.*

necessary because state-of-the-art FPGA chips currently come in packages with over 100 pins in closely packed configurations; manual soldering of such chips is impractical, and the fabrication of printed circuit boards for individual FPGA designs is costly and time-consuming. Even with pre-made FPGA prototyping boards, the students should still learn the entire FPGA configuration process, including the configuration file download procedure, and use this knowledge to use the FPGA prototyping board properly. The FPGA prototyping board should simply provide the wire connections required to enable configuration using externally provided header pins. This paper presents the design and philosophy behind one such custom-designed FPGA prototyping board. Based on the method and design presented in this paper, it is also possible to design other FPGA prototyping boards for other types of FPGA chips.

2.   FPGA CONFIGURATION

An FPGA (field-programmable gate array), as its name implies, can be programmed in the field (i.e., by the end user). The overall FPGA configuration (also referred to as programming) process is as follows. First, the user must create his/her design using software EDA CAD tools (such as Exsedia's Nimbus) and then convert it into a "configuration" file format. Then the configuration file must somehow be downloaded to the FPGA chip. Control circuitry within the FPGA then takes this data and uses it to create or break connections at predetermined locations (typically horizontal-to-vertical crosspoints) dispersed throughout the chip. The circuitry within an FPGA chip is typically laid out in a rectangular two-dimensional matrix of configurable logic blocks (cells) separated by intervening rows and columns of wires (the routing channels). By creating customized connections, cells can be configured (or programmed) to implement specific subcircuits (such as 2-bit adders or counters) and connections can be made to the wires in the routing channels to connect cells in different locations.

As an example of an FPGA and the FPGA configuration process, let us examine the Spartan II XC2S200 FPGA produced by Xilinx. This is a 2.5 volt CMOS device with the equivalent of 200,000 logic gates, a 56-Kbit RAM memory, and 140-284 user I/O pins. As with most Xilinx FPGAs, it uses SRAM cells to store the configuration information, and is thus a volatile device that can undergo an almost unlimited number of reprogramming cycles. This chip supports the slave serial, master serial, slave parallel, and boundary scan modes of configuration, with the specific configuration mode selected by setting the values of three mode pins M0-M2. All configuration modes supported are synchronous modes, in which a separate clock signal (CCLK) is used to indicate the presence of valid configuration data on the data input (DIN) line. In the slave serial (or parallel) mode, the CCLK and DIN (or D0-D7) signals are generated externally and used to download the configuration data to the FPGA. In the master serial mode, the FPGA generates the CCLK signal and reads in the DIN signal at the appropriate times. Finally, the boundary scan mode "reuses" the I/O pins provided for the production testing of this device, referred to as the JTAG access port, in order to download configuration data in a slave serial-like manner.

3.   DESIGN OF THE FPGA PROTOTYPING BOARD

The FPGA prototyping board was designed to facilitate experimentation with numerous different types of digital logic circuit designs. Since it was not designed to be a commercial product, software support was not provided. Instead, we simply relied on the configuration software available with most FPGA tools to download the configuration data to our board.

Figure 1 shows a block diagram of the FPGA prototyping board that was produced. There are several notable points about this board design. First, all user I/O pins are brought out to header pins so that they can be probed easily using logic analyzer or oscilloscope probes. If we wish to connect this board to another circuit board and drive inputs to the FPGA from the external board, that can be easily accomplished by connecting to the header pins provided.

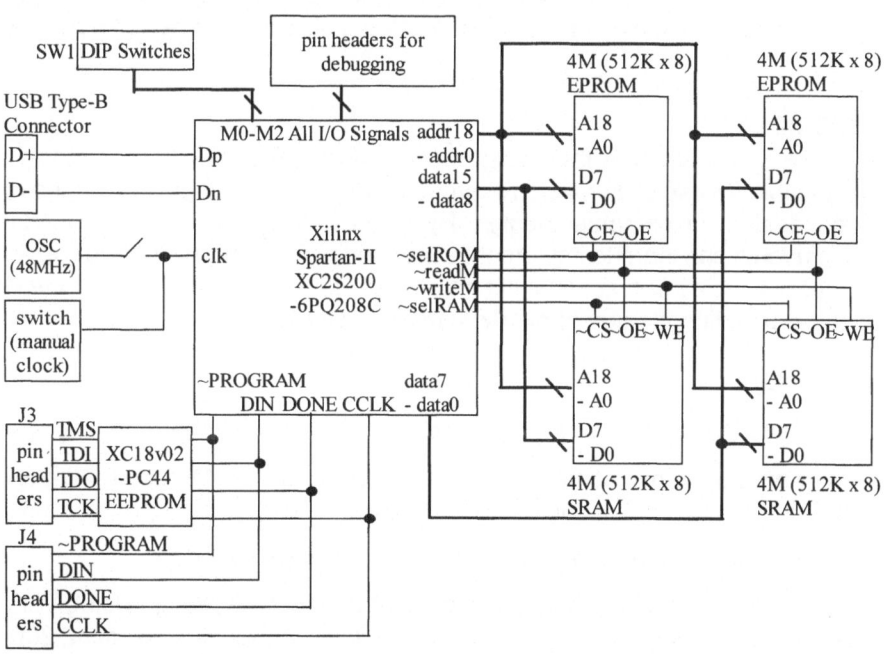

Fig. 1: Block diagram of a custom-designed FPGA prototyping board for a Xilinx Spartan II FPGA

As can be seen from Figure 1, connections have been provided for configura-tion using both slave serial mode and boundary scan mode. The specific mode used can be selected by connecting the mode pins (M0 – M2) to ground or Vdd. In particular, the boundary scan configuration pins have been connected to a serial EEPROM. Thus, configuration data can first be downloaded to the serial EEPROM using boundary scan mode. Then, since the EEPROM is a non-volatile device, power can be removed, the mode changed to master serial mode, and the configuration data downloaded to the FPGA at any time after the power is restored. The FPGA can also be configured directly using slave serial mode.

Since we wished to be able to experiment with modern I/O interfaces and memory devices, we also added connections to a USB (universal serial bus) port and connections to SRAM and EPROM memory devices. The USB I/O interface was selected because USB Version 1.1 can be implemented entirely in digital logic (with no analog components) and because USB is a popular modern, high-speed, serial interface representative of the current trend toward high-speed serial I/O interfaces. A 48MHz oscillator chip was added in order to enable operation of the USB interface at 12Mbps. [4] provides information on example Verilog and VHDL designs for a USB 1.1 protocol analyzer using this interface. SRAM and EPROM devices were added to enable experimentation with circuits requiring external memory, such as DSP and general-purpose microprocessor designs. In a class on "Computer Design" at our university, students were able to implement a pipelined RISC microprocessor for the ARM THUMB instruction set using this circuitry [4].

DISCUSSION

This paper has presented the design of a general-purpose FPGA prototyping board suitable for use in a university course on digital logic design, advanced digital logic design, or computer design. By using this type of FPGA prototyping board, students can learn to implement advanced digital designs, possibly involving memory devices or fast serial I/O interfaces, using state-of-the-art FPGA devices while avoiding the use of overly "canned" solutions.

REFERENCES

[1]   www.synopsis.com, home page for Synopsis Inc., a synthesis tool vendor.

[2]   www.exsedia.com, home page for Exsedia, an ASM-based high-level design capture and simulation tool vendor.

[3]   www.xilinx.com, home page for Xilinx Inc., an FPGA vendor.

[4]   S. Lee, Advanced Digital Logic Design: State Machine Design Using VHDL, Verilog, and Synthesis for FPGAs, Brooks/Cole Publishing, Belmont, 2004.

DESIGN OF PROTOTYPING BOARDS WITH XILINX FPGAS

LEYVA G., CAFFARENA G., CARRERAS C., NIETO-TALADRIZ O.
Universidad Politécnica de Madrid, ETSIT – DIE, Av. Universitaria s/n, 28040 Madrid, Spain.

ABSTRACT
In this paper we propose a method to develop low-cost training FPGA-based boards for fast prototyping of digital circuits aiming ease of use. This method allows the implementation of prototyping boards targeting all Xilinx FPGA families including SPARTAN-III and VIRTEX-II PRO. The board configuration is carried out in serial mode (master and slave) through a microcontroller, although JTAG is also supported by some of our final designs. RS-232 serial mode configuration makes board programming independent on the OS platform in use. Daisy-chain configuration is feasible for those designs that require multiple FPGAs interconnected.
We present examples of different training boards intended for academic laboratories that support a wide range of SPARTAN, VIRTEX and VIRTEX-E FPGA devices. The implementation cost of the boards is in average a 50% of the cost of alike commercial products.

1. INTRODUCTION

The goal of this paper is to present a method to develop FPGA-based training boards for fast prototyping of digital circuits. The method seeks low-cost and easy to use final designs. The proposed training boards conform a platform for HDL-based digital design, targeting students or professors that would like to investigate the applications of Xilinx's FPGAs [1] under low resources constraints.
The training boards contain a low-cost microcontroller AT89C2051 [2] that acts as an interface between the computer and the FPGA. The download process can be carried out by any serial communications software through the RS-232 port at a speed of 57600 bauds.
The programming method proposed has been applied to two training board models: a) based on XC4000 and SPARTAN families (up to 20K gates) and b) based on VIRTEX and VIRTEX-E families (up to 400K gates). The former provides 24 I/O and additional hardware (7-segment display, push buttons and dipswitches) and has an approximate cost of 65 €. The latter provides with 150 I/O and has an approximate cost of 200 €.
It is important to stress that the same implementation method can be applied to more recent FPGAs as VIRTEX-II/PRO [3] and SPARTAN-III [1].
The good results obtained allowed us to start up a Digital Design Laboratory for 200 students with 30 low-cost boards. On one hand, the use of slave serial configuration saved a big deal of money since the Xilinx JTAG configuration cables

A.M. Ionescu et al. (eds.), Microelectronics Education, 81–85.

[4] were not necessary. On the other hand, serial configuration works in both UNIX and Windows platforms through the RS-232 port, while JTAG parallel port configuration is not compatible with all systems.

2.   BOARD DESCRIPTION

The main parts of the board are: Xilinx FPGA, microcontroller, EEPROM, RS-232 interface chip, status LEDs and crystal oscillator.
The design bitstream is necessary in order to program the boards and it must be generated by the Xilinx Development System [5]. The microcontroller interprets the design bitstream, configures the FPGA and monitors configuration errors and activity. The microcontroller, FPGA device and EEPROM are connected via a serial bus. This serial bus enables the connection to other board's FPGAs through serial configuration ports as well.

2.1   INTEL MCS86 BITSTREAM

The MCS86 format was selected for the configuration file that the boards support. It can be generated by all versions of Xilinx Development System. The MCS86 format is analyzed by the microcontroller and, thus, it is possible to detect any communication and configuration errors that may occur during the bitstream download process.

2.2   MICROCONTROLLER

The microcontroller program was written in assembler for the sake of efficiency and it carries out the following tasks: design bitstream reception, board status monitoring, EEPROM programming, FPGA configuring and clearing.

2.3   EEPROM FOR MASTER SERIAL MODE

The FPGA and EEPROM [6] bus is mainly conformed by a clock signal and a bidirectional data signal. The selected memory for the SPARTAN-based board is the AT17LV128, and the memory for the VIRTEX/E-based board is the AT17LV010. In both cases the write operation is performed by using 64-byte and 128-byte pages, respectively.

2.4   MONITORING

An interesting feature of the boards is that users can monitor the board status by means of a set of LEDs. These LEDs operate continuously. The following states are considered: Ready, Error, Rx and Done.

3.   FPGA PROGRAMMING

Xilinx's FPGAs can be configured using several modes. Even although JTAG is widely used for prototyping, we make use of the serial slave mode's features. This mode is supported by all Xilinx FPGA families [1].

The microcontroller provides signals CLK and DIN, which are necessary to con-figure the FPGA, as well as signals PROG and INIT. During configuration, FPGA signal INIT is continuously monitored to assure the correct bitstream download.

Each board has both input and output configuration serial ports for daisy-chain configuration. Thus, if it is necessary to segment a design that requires several FPGAs for its implementation, the board can be daisy-chained and all boards can be configured via a single code download and using a single RS-232 connection to the computer. The Xilinx Development System is able to obtain a single config-uration file for such cases.

Figure 1 displays a photograph of the training board [7] based on XC4000 and SPARTAN families. Any FPGA from those two families can be used as long as the package is PLCC84 and the number of equivalent gates does not exceed 20K.

Fig. 1: FPGA training board based on Spartan and XC4000 families

Figure 2 shows a photograph of the board based on VIRTEX and VIRTEX-E fam-ilies. This board allows the use of JTAG configuration mode as well as the serial mode previously described. Any FPGA from the two families mentioned can be fit into the board as long as the package is TQFP 240 and the number of the equiv-alent gates of the device does not exceed 400K.

The configuration method described in this section also applies to more recent Xilinx FPGA's families as VIRTEX-II/PRO and SPARTAN-III.

CONCLUSIONS

The results presented here clearly show that the implementation of low-cost train-ing boards to be used at University FPGA Design Laboratories is a viable option. Despite the fact that there are numerous commercial training boards with high-density FPGA devices [8,9,10,11], we consider that our low-cost approach offers a more flexible alternative for low/medium density digital systems prototyping. The boards presented achieve a 50% cost reduction compared to commercial products.

The use of JTAG configuration cable is not necessary, leading to further cost savings. However, this cable can be used in the VIRTEX-based board. The fact that the board can be configured via the RS-232 port makes possible to program the board in Linux and Windows O/S, resolving the problematic surrounfing teh use of the parallel port in different platforms.

Fig. 2: Training FPGA board based on VIRTEX and VIRTEX-E families

Our approach allows the development of prototyping boards targeting all Xilinx FPGA families, including SPARTAN-III and VIRTEX-II PRO. It is possible to daisy-chain several boards during the configuration process for designs that requires several FPGA devices.

The method presented in this paper has been applied to the design of several Xilinx FPGA-based training boards. These boards have been successfully introduced into digital design university laboratories .

REFERENCES

[1] Xilinx Databook DS099-2(V1.2), pp.32-39, July 2003.

[2] http://www.atmel.com/dyn/resources/prod_documents/DOC0368.PDF

[3] Xilinx VIRTEX-II PRO, Platform FPGA Handbook, pp. 68-71, 2002

[4] http://www.toolbox.xilinx.com/docsan/xinilx4/data/docs/pac/cables6.html

[5] http://toolbox.xilinx.com/docsan/xilinx6/books/data/docs/dev/
dev0126_19.html

[6] http://www.atmel.com/dyn/resources/prod_documents/doc2321.pdf

[7] G. Leyva, R. Jaramillo, "Tarjeta didáctica para prácticas de circuitos
digitales en FPGA", Proc. Computación Reconfigurable & FPGA's, UPS
Universidad Autónoma de Madrid (Spain), Eds. Boemo, Gómez, López-
Buedo, Sutter,pp. 457-462, September 2003.

[8] www.avnet.com

[9] www.associatedpro.com/v240revb.PDF

[10] http://www.cesys.com/english/ebene2/productp.htm

[11] http://www.xess.com/prod032.php3

SW/HW INTEGRATION IN EMBEDDED SYSTEMS DEVELOPMENT

A Teaching Methodology
RA O., STROM T.
BusKerud University College, P.B.251, N-3601, Norway

1.   INTRODUCTION

The rapid expansion of the "Embedded Systems" segment in general, and the "System-on-Chip" (SoC) area in particular, is creating many new challenges at all stages of the design process. It is generally acknowledged that adoption of a suitable process development methodology, unified with respect to software and hardware system constituents, will become mandatory in order to improve productivity, thus enabling harvesting of the potentially vast design spaces offered by ever increasing advances in semiconductor technology.

2.   RATIONALE

In our experience there is a severe bottleneck in the design of an effective curriculum for embedded systems. Many universities and colleges have sets of courses which approach several of the important problems in design and development of embedded systems. However, there seems to be a serious lack, with respect to *cohesion, organization and packaging.*

There is not enough time to select all pertinent themes from a standard university/ college menu and then to study them in depth (the *organization and packaging problem).* Course modules typically reflect different hardware and software cultures (the c*ohesion problem*) so a simple downscaling of theme volumes does not provide a satisfactory solution.

Traditional SW and HW development methodologies/techniques/tools have scarcely been able to keep track of exploding complexity in the separate domains. Future embedded systems development methodologies/techniques/tools even have to become integrated to tackle aggravated development problems.

3.   OUR APPROACH

This presentation describes our attempt at creating a unified and integrated approach where a systems focus is key and where SW and HW topics are taught concurrently with SW and HW modules being "aware" of each other.

A unified methodology stands to profit by an object-oriented and unified-system-description-capability offering scope for validation through executable specifications, for verification during gradual refinement, and a starting point

A.M. Ionescu et al. (eds.), Microelectronics Education, 87–89.
© 2004 *Kluwer Academic Publishers.*

for software code generation as well as communication and digital hardware synthesis.

Hence we devote a major part of the presentation to teaching experiments with a *Rational Unified Process (RUP)* type development process, suitably modified to account for HW and also HW/HW, and SW/HW communication models.

The model exploits **UML** as a **co-modeling** tool to obtain:

- *A common, structured environment for documentation of requirements and design*
- Enhanced inter-disciplinary communication – e.g. sequence diagrams showing complex SW/HW interactions to reduce misunderstandings across domain borders
- Useful separation of abstraction levels to aid understanding

and makes use of **SystemC** to obtain a multi-level **executable** layer with:

- Explicit mapping of stereotyped UML constructs onto SystemC concepts

Recent guidelines for bridging the mindsets of the SW and HW worlds, along with input to weighting and balancing of technical curriculum constituents have been derived from our work in the "CoDeVer" project; financed by the *Norwegian Research Council* and focused on methodology, tools and best practices for the co-design and co-verification of embedded systems:
http://www.ittf.no/prosjekter/codever

Our approach has been used with our own students for two years, including internal assignments as well as projects in cooperation with industry. Some projects comprise complete, iterative and object oriented development cycles from requirements gathering and specification, through analysis and design, down to implementation with mapping of HW on FPGAs carrying embedded processor cores

4.   FURTHER DEVELOPMENT

To retain industrial relevance in its teaching approach HiBu carries the "CoDeVer" effort further in the "EMBLA" project, also financed by the *Norwegian Research Council*,
http://www.abelia-innovasjon.no/page/main.php?top=3&content=18&fg=792&instance=792
and is responsible for planning and progress in a strand devoted to systems modeling. The presentation will comprise a report on results from cooperation with our industrial partners, which results are continually fed back into our curriculum and teaching practices.

Our methodology and practice has recently, within the "Soc-SME" project under the auspices of *The Nordic Industrial Council*, been implemented also in a distance-learning, web-based course format aimed at engineers already working industrial companies; see:

http://www.abelia-innovasjon.no/page/main.php?top=3&content=18&fg=775&in-stance=775

Feedback from test runs with trialists from industrial companies in the *Nordic and the Baltic* regions will be analyzed in time for inclusion in the presentation.

TRAINING FOR SOC DESIGNERS: TODAY'S SOLUTIONS FOR TOMORROW'S PROBLEMS?

*STECHELE W. **DE MEY B.
*Technische Universität München, D-80290 München, Germany,
Walter.Stechele@ei.tum.de, **IMEC, Kapeldreef 75, B-3001 Leuven, Belgium

Abstract

Growing complexity of System-on-Chip (SoC) designs, faster development cycles, and increasing power restrictions are challenging factors for design engineers. Competing with these challenges requires continous training on advanced design techniques. This paper deals with the selection of training course topics from a course provider's perspective. A course on low power optimization for integrated system design is taken as an example on how to map controversial industrial needs for training on advanced course topics. Industry has to solve today's problems and invest in solutions for tomorrow's problems at the same time with limited budget. These controversial requirements are not sufficiently solved today.

Keywords

System-on-Chip design, low power, training

1.   Introduction

Design engineers for complex Systems-on-Chip (SoC), that include multiple processor cores, memory systems, bus architectures, and lots of software, are facing growing challenges in order to compete with the design productivity gap. This requires continuing education to keep up with the latest, most advanced design tools and methodologies. A European consortium under the guidance of ECSI [1] has established a training activity for system-level electronics design. The goal is to provide training courses on advanced system design for engineers from industry, where no commercial courses are available. This activity is supported by the European Commission in the project SYDIC-Training [2]. The consortium consists of IMEC, ISLI, LIU, TUM as research institutes providing training, Cadence, Mentor as EDA training providers, and ARM, Ericsson, MBDA, Motorola, as industrial partners providing input to the definition of course content and delivery style, i.e. the need for training. The project started in 2002 and continues till 2004, with phase 1 concentrating on course content, phase 2 concentrating on delivery mechanisms. First courses started in 2002 and are planned to continue within the framework of ECSI after the project.

The university partners are providing courses on advanced topics on SoC design that transfer recent research results to industry.

A.M. Ionescu et al. (eds.), Microelectronics Education, 91–94.

This paper analyses the process of selection of topics for advanced system design training courses. In section 2 relevant criteria are shown, which have been used to define course content. In section 3 the course on low power optimization from TUM [3] is taken as an example to show, how we tried to fulfill the controversial requirements on course content. In section 4 the criteria for content selection are mapped against the content of the course and conclusions are drawn.

2.   Selection of Course Topics

The criteria for selection of course topics, which have been put together in the SY-DIC-Training project, are quite controversial.
- Advanced
- Applicable today
- Available tools
- Relevant

Topics have to be advanced. I.e. they have to go beyond state-of-the-art design methodologies or tools. At the same time, they should be applicable today, with available EDA tools that are ready-to-use.

Topics have to target relevant problems. This seems obvious, but considering relevant problems of today requires instant solutions, which makes it too late for training.

Other requirements are even more controversial. Some designers are looking for application-specific courses, some require unspecific, purely methodology-oriented courses. Some are looking for tool-oriented courses - but only if the tool is used in the company, others generally require non-tool-specific courses.

In the following section, the course on low power optimization for integrated system design is briefly described, in order to analyze how the various topics of this course are mapped on the above shown criteria for selection of course topics.

3.   Low Power Optimization for Integrated System Design

The goal of the course is to give an overview over a broad range of low power techniques on all levels of digital circuit design. Then concentrate on selected techniques: (1) Clock gating, (2) Dual Supply Voltage Scaling, and (3) Register Transfer Level Power estimation. All techniques are applied using standard EDA tools.

(1) Clock gating

Clock gating is a technique to reduce dynamic power by switching off clock signals at those blocks of the circuit, where no useful activity occurs. This is done thru extra logic to analyse signal activities and generate clock gating control signals.

(2) Dual Supply Voltage Scaling, DSVS

DSVS uses two levels of supply voltage (Vdd) in logic synthesis [5]. All cells in the synthesis library might be used with high Vdd (for high speed, high power) or low Vdd (for lower speed, lower power) respectivly. The selection of high/low Vdd cells is done automatically by the synthesis tool (Synopsys Power Compiler).

Level converting Flip-Flops have been included into the design library and connection classes have been introduced to automatically control the transitions between high/low Vdd domains.

With this technique, timing slack is used to minimize power, in addition to all conventional optimization techniques used by Power Compiler.

(3) Register Transfer Level Power estimation, RTL-Pest

Low power optimization can be supported by Register Level Power estimation, which is a technique to simulate power consumption of register transfer level models, based on signal activities stimulated by a testbench [6].

Various techniques for modeling power consumption, on bit level and word level, as well as characterization of RTL modules, are presented and discussed in the course.

In the next section, the three dedicated techniques of the low power course, i.e. (1) clock gating, (2) DSVS, (3) RTL-Pest are mapped against the criteria for course content selection.

Conclusion

Table 1 summarizes how the criteria for the selection of course content are met by the dedicated techniques of the low power course.

Criteria	Advanced	Applicable today	Available tools	Relevant topic
(1) clock gating	≈	✓	✓	✓
(2) DSVS	✓	≈	≈	✓
(3) RTL-Pest	✓	✓	✓	✓

Tab. 1: Course content of low power course versus criteria for content selection
(☉ high, ≈ medium)

The detailed analysis for the course topics is as follows:

(1) Clock gating is medium level advanced, as it is applied for many years manually in RTL code and supported by EDA tools, but its potential is not fully exploited by available tools.

(2) DSVS is highly advanced, as it uses voltage scaling in addition to conventional optimization techniques (transistor scaling, pin swapping, and others). It can be applied today, but voltage level converting Flip-Flops have to be developed and included in the design library before. It can be used with standard synthesis tools, e.g. Synopsys Power Compiler, but available layout tools for two-level Vdd ´, e.g. CATENA [7], are available today only for 2-layer metal, and need to be adapted for multilayer metal.

(3) RTL-Pest is highly advanced power estimation on RT level, based on signal activity from test bench, faster than gate level simulation, but with similar accuracy. It is applicable today with appropriate RT module characterization. Standard simulation tools are used.

As it can easily be seen, the course on low power optimization for integrated system design offers a mixture between ready-to-use techniques and more advanced techniques, which might be used not today, but tomorrow. Solving tomorrow's problems requires investing in advanced solutions today.

Industry's focus today is to invest in the building of skills to solve its today's problems. Its short time focus however can lead to insufficient investment in the building of skills it will need to cope with the growing design productivity gap.

Acknowledgment

The authors want to thank Paul Zuber from Technische Universität München for his contributions to the low power course.

This material is based upon work supported by the IST programme of the EU in the project IST-2001-35100 SYDIC-Training.

References

[1]   www.ECSI.org

[2]   IST project SYDIC-Training, IST-2001-35100,
       www.ECSI.org -> SYDIC-Training

[3]   www.lis.ei.tum.de

[4]   http://www.lis.ei.tum.de/sydic/indexlp.html

[5]   Mahnke, T., Stechele, W., Höld, W.: „Dual Supply Voltage Scaling in a Conventional Power-Driven Logic Synthesis Environment", International Workshop on Power and Timing Modeling, Optimization and Simulation, PATMOS 2002, Sevilla, Spain, September 11.-13. 2002

[6]   Eiermann, M., Stechele, W.: „Novel Modeling Techniques for RTL Power Estimation", International Symposium on Low Power Electronics and Devices, ISLPED 2002, Monterey, CA, August 12.-14, 2002

[7]   www.catena-ffo.de

SELF-INSPIRED LEARNING ASSISTED BY EDA TOOLS IN THE INTRODUCTORY DIGITAL DESIGN COURSE

LINGHAO S., BURIAN A., TAKALA J.

Institute of Digital and Computer Systems, Tampere University of Technology
P.O. Box 553, FIN-33101 Tampere, Finland.
Fax: +358 3 31154561. E-mail: {linghao.sun, adrian.burian, jarmo.takala}@tut.fi

1.   INTRODUCTION

The Institute of Digital and Computer Systems is an independent research unit of Tampere University of Technology, Finland. The institute is in charge of numerous research projects, as well as both under-graduate and post-graduate teaching activities related to digital systemic and computer architectural design. The rapid growth of the number of international students, coming from both developing countries and advanced ones, with different educational background and levels of knowledge, imposes a true challenge on the teacher, especially for introductory level courses.

The goal of this paper is to present our experience in introducing industrial Electrical Design Automation (EDA) tools in the teaching process of basic courses, with the "Basic Digital Design" course as target. With EDA tools, the learners have a virtual experimental environment, and are able to test and experiment what they have learned in the lectures. Furthermore, they are able to practice the digital design challenges. This type of "non-paper work" assignments makes students more interested in their otherwise considered plain homework, and therefore a self-inspired and self-motivated study atmosphere is established. The students will then be well prepared for further studying advanced microelectronic systems. In addition, we discuss the merits of using the scientific method and a cooperative learning methodology for basic courses, with the particular example of our course.

2.   A DOORSTEP TOWARDS A QUALIFIED ELECTRICAL ENGINEER

In our institute, courses like "Basic Digital Design" or "Computer Architecture" are introductory level curricula material for students with their major subject digital and computer systems. The follow-up courses include "Processor Design", "Signal Processors", "Digital ASIC Design", and "Design on Silicon". The period of study dedicated to basic digital concepts is transitional, when the student will step forward from their study of fundamental and pure theoretical curricula towards more practical and realistic ones. The firmness of this doorstep is significant for the students' involvement in advanced courses, and also important for the advancement of students towards electrical engineers.

A.M. Ionescu et al. (eds.), Microelectronics Education, 95–99.

When stepping into this study period, the learner's custom is to be concentrated on the results rather than on the procedures. This is because mainly the correctness of their results reflected their exam marks. It is no doubt that this type of study cannot be neglected during the preliminary study period, but with the development of the knowledge, this method of learning will limit the study characteristics. A good "result" of a digital system design will not only depend on a good principle, but also on the cost, time-to-market restriction, possibility of realization, system working stability, and even the capability for quality control, manufacturing management etc., and all of these items demand reasonable and intelligent design *procedures*. Thus, a role switching is needed for the learner in order to become a qualified electrical engineer.

Traditionally, homework assignment by paper work is a typical way to review the knowledge from classroom. The biggest drawback of this kind of assignment is the difficulty for the learners to proof the design results and to justify the correctness of their specific design by themselves. Laboratory assignment is another efficient method to overcome the self-justification obstacle, and offers a good way to experience what the real-world digital design is. Its main weakness is the practical limitation given by the available type of elements – devices and gates. Both previous methods can be applied at the cost of extra time needed for introductory curricula and increased expenses for laboratory hardware. By using EDA tools, these additional costs (time and money) are avoided [1-3, 7]. We mention that an alternative approach can be the usage of Java-applets (interactive modules) running on any browser connected to Internet ("living pictures") [5, 6]. There are some limitations to this approach at the technical level. Another interesting idea is to use programmable devices, like in [4]. Using a programmable approach implies the academic curriculum to be revised and a risk investing in the future.

3. REALIZING BOTH THEORY AND EXPERIMENT BASED ON EDA

The EDA tools have been developing fast, and they became beneficial for both academic and industry users. Today's EDA tools are no longer a pure simulation alternative; they involve the user in all the design procedures of digital and computer systems.

From educational point of view, EDA tools have several advantages. Firstly, the design procedures are visualized and transparent to the user. Secondly, the design process becomes an interactive procedure by easily changing the value of parameters. Some advanced EDA tools allow the user to create new library or element models. This kind of multi-level implementation and thinking is impossible for a student in the laboratory works, meanwhile, the transparency of each level or layer makes it easier to learn and observe.

The third advantage of EDA tools is that simulation as well as its related analyzers gives the user the possibility of self verification and justification. This is one major change from a conventional approach, where the teacher must be responsible for both making assignment and justifying the solutions of the students. With EDA tools, the students may not have to wait the verification from the teacher, and are able to justify their solutions by themselves. The forth advantage is the high flexibility of the design procedure. This is significant to bring up the creativity of the students as well as to illustrate the need for "procedures".

4.   INITIATIVE STUDY IS THE KEY FOR INTRODUCTORY COURSES

To be the doorstep of advanced courses, introductory courses of digital and computer systems still can not avoid some plain characteristics, even if they feel more close to real world than the preliminary courses. This is because the curricula still emphases the basic principles, properties and definitions rather than a design skill training course. For our international and various study background students we have been using two different approaches. At first, we arranged the course to be formed from lectures and exercise sessions for homework verification (paper & pen). After 3-4 sessions of exercise works, the students start loosing interest, since nothing new compared to their previous studies appeared. We had this experience in our first years of teaching "Basic Digital Design". When no practical works were assigned, usually the marks decreased from "5" (highest), "4" at the beginning of the period to "1" or "0" (no returned solution) in the later assignments. This deviant distribution clearly reflected the attitude of the students. On the other hand, for students with some background knowledge of digital design, the assignments were not enough to supply their demands and draw much interest. In the latter years, we introduced cooperative learning using the EDA tools. This allows changing the student's role from a knowledge "receiver" into a knowledge "verifier". Since we adopted the "non-paper work" and computer based assignments, more debate concerning the digital design methodology or optimization during or after lectures and exercises between the teacher and the students has been taking place.

In Figure 1 we illustrated through some charts the result of engaging cooperative learning in teaching a first course in digital design, and its overall impact. The students' opinion about this learning method appears clearly from their responses. We mention that the questionnaire has been completed anonymously. It is noticeable that more than half of responses declare their overall satisfaction to be more than 90%.

REFERENCES

[1]   N.L.V. Calazans, F.G. Moraes, "Integrating the Teaching of Computer Organization and Architecture with Digital Hardware Design Early in Undergraduate Courses", *IEEE Transactions on Education*, vol. 44, no. 2, May 2001, pp. 109-119.

[2]   D.V. Hall, "Teaching Digital Methodology and Industrial Strength EDA Tools in a First-Term Freshman Digital Logic Course", *IEEE Transactions on Education*, vol. 41, no. 1, Feb. 1998, pp. 45-50.

[3]   R.D. Meier, C.T. Wright, Jr., "The Use of Simulation Based Laboratory Exercises to Improve the Undergraduate Digital Logic Design Experience", *IEEE Frontiers in Education Proceedings*, vol. 3, Nov. 1996, pp. 1177-1180.

[4]   M.S. Nixon, "On a Programmable Approach to Introducing Digital Design", *IEEE Transactions on Education*, vol. 40, no. 3, Aug. 1997, pp. 195-205.

[5]   R. Ubar, E. Orasson, H.D. Wuttke, "Internet-Based Software for Teaching Test of Digital Circuits", 23rd *International Conference on Microelectronics (MIEL 2002)*, vol. 2, May. 2002, Nis, Yugoslavia, pp. 659-662.

[6]   H.D. Wuttke, K. Henke, "Teaching Digital Design with Tool-Oriented Leaning Modules Living Pictures", 32nd *ASEE/IEEE Frontiers in Education Conference*, Nov. 2002, Boston, MA, pp. S4G25-29

[7]   Yong Y. Li, "Experiences Teaching Design Automation in the Introductory Level Course", *IEEE International Conference on Microelectronic Systems Education*, July 1997, pp. 64-65.

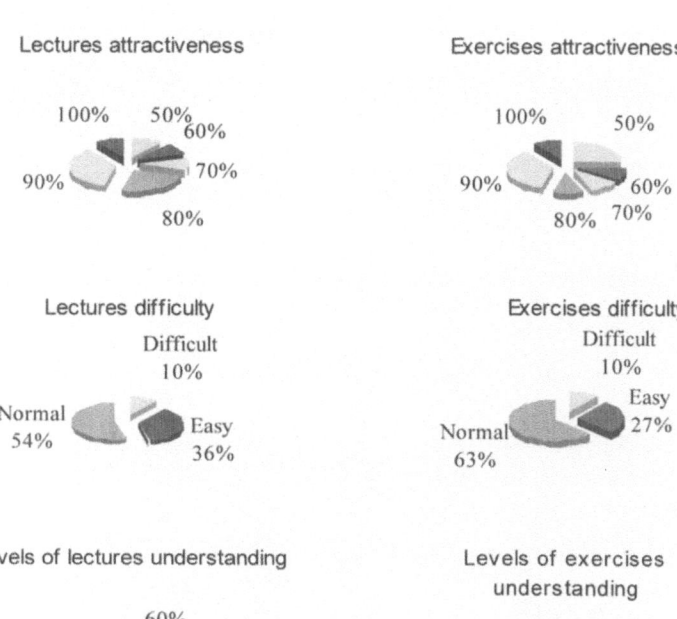

Lectures attractiveness

Exercises attractiveness

Lectures difficulty

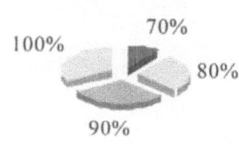

Exercises difficulty

Levels of lectures understanding

Levels of exercises understanding

Levels of overall satisfaction

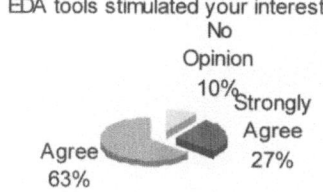

Extra weekly hours self-studying

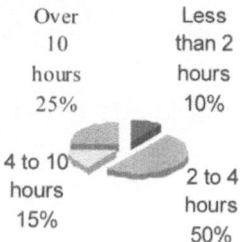

EDA tools stimulated your interest

EDA tools are valuable supplement to lectures

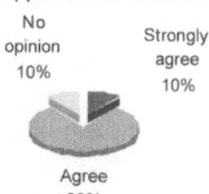

Fig. 1: Summary of questionnaire responses

SUCCESSFULLY GETTING IN TOUCH WITH FORMAL DESCRIPTION LANGUAGES

An XML-based practical training on VHDL
WALDHAUER M., LUCKE U., TAVANGARIAN D.
University of Rostock, Dpt. of Computer Science, Chair for Computer Architecture, Albert-Einstein-Str. 21, D - 18059 Rostock, Germany

1. THE NEED FOR INNOVATIVE eLEARNING MATERIAL

Currently, teaching on the one hand and exercises or practical training on the other hand are often just loosely coupled. Especially, environments for practical training (if implemented in electronical manner) sometimes are not very comfortable, are putting students off rather helping them, nor allow teachers for examination of student's activities apart from the mere experimental results. Therefore, from a didactical point of view, the success of these arrangements is likely to be questioned. This is especially true for abstract subjects like formal description languages and their practical application.

Moreover, new learning scenarios (e.g. mobile learning, blended learning at a notebook university) are requiring more powerful educational material and tools [1]. Basic description languages (at this moment these are usually HTML or Java) will experience a fundamental change: There is a strong demand for highest flexibility according to a large number of parameters, resulting in documents with various instances for different situations. This becomes very complicated to handle if dynamic multimedia or interactive elements are used.

To solve such problems this paper introduces a scalable model to meet with these difficulties, an XML-based language that implements the model, and it's application for the development of a practical training on the hardware description language VHDL, and discusses some aspects using this model in the sense of teaching & learning at a university.

2. MULTIDIMENSIONAL SCALABILITY OF eLEARNING MATERIAL

2.1 THEORETICAL MODEL

Conventional educational material is more or less a static object, classified by the author for use in a specific scenario. But wide re-usability requires the possibility to adjust some characteristics of the object by the user (teacher or learner) regarding to their individual situation [2]. This includes:

- *learner type*, e.g. pre-knowledge, language, learning style (e.g. behavioural, instructive or constructive), gender, and so on;

A.M. Ionescu et al. (eds.), Microelectronics Education, 101–105.
© 2004 *Kluwer Academic Publishers.*

- *learning environment*, e.g. available amount of time, availability of a teacher or tutor (or self-learning), required file type, or simply the style (corporate identity);
- *learning goals*, e.g. detailed list of subjects, educational context (e.g. other material), educational forms (theory or practice), or an overall didactic strategy.

All single aspects are more or less independent and can therefore be interpreted as dimensions of a learning object. They need to be modelled by the author of an object, which results in a large number of possible instances created from a single, abstract description. But multi-dimensional scalability as described above is hard to implement. Continuous co-domains are hard to discover and evaluate. The authoring process for a multi-dimensional object can be assumed as too complex, as well as the effort for transformation and management of different instances at run-time. For these reasons, a restriction to essential aspects and discrete co-domains with limited values is necessary.

2.2 THE <ML>³ LANGUAGE

The XML-based description language <ML>³ (Multidimensional LearningObjects and Modular Lectures Markup Language) implements this theory in a simplified way [3]. It is based on a three-dimensional model, considering educational intensity (for basic, advanced or expert users), target person (student or teacher) and output device (as online (xHTML), slide (SVG, PPT) or manuscript (PDF) version). Other output formats would be possible (e.g. WAP), but not always feasible. Additionally, various didactical structures can be described for a learning object. These structures are detached from the content itself, and are exchangeable.

The language is described in four XML schemata, defining elements and their use conditions. Scalability of a learning object is implemented as attributes of its elements, which are evaluated by appropriate output processors. They generate up to 18 different instances of a learning object, according to the possible values of each of the three dimensions. Furthermore, individual re-arrangement of an object is flexibly done by changing its didactical structure. Several learning objects can be linked among each other (or with external sources), and can be combined to courses. They are accessible via a publication platform. A variety of additional <ML>³ tools is available.

3. PRACTICAL TRAINING ON VHDL

Since years there is an apparent development in the area of training on VHDL using electronic ways for presentation of the learning material and providing platforms for doing the practical training, which in several cases led to successful products or online courses like [4], [5], [6], [7], [8].

The new opportunities provided by <ML>³ are the basis for a large set of eLearning material, which covers most topics of computer engineering. This material has been split into about 150 so called modules, among them synthesis of digital circuits using VHDL. The modules usually belong to one of the categories theory or practice.

Because of different requirements for practice modules, their structure is different than for theory modules. It is also an effect of some topics (e.g. VHDL synthesis) being too rich in content to fit into a single or two modules. The implementation of such a practice module needs the formulation and transformation of exercises and belonging material (introductions, hints, descriptions, etc.), and also the preparation of experiments or a development platform, where the users are able to do experiments or develop and test the solutions of given tasks.

We provide centralized simulation or experiment platforms, thus standardizing the use of the tools and offering the ability to connect the practice works with the module. Experiments and simulators are included by inserting interactive elements (e.g. Java applets, links to external web pages, interactively solvable tasks) into a module, which give remote access to them [9]. For the VHDL practice modules this is the VHDL simulator, which is located on a central web server. In this way these modules offer a consistent integration of development and simulation into the learning process via application service providing.

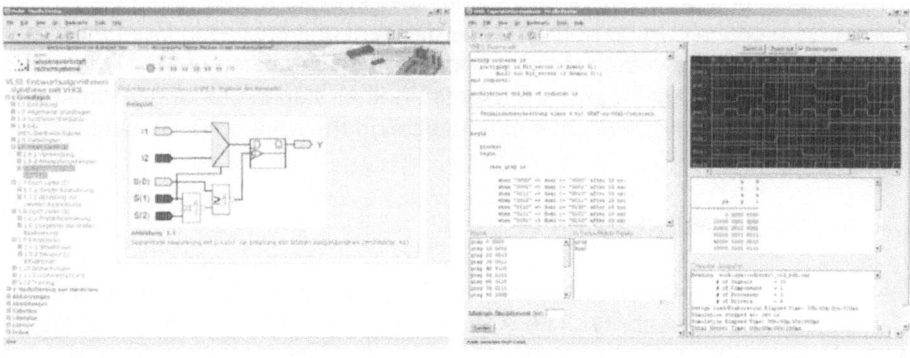

Fig. 1: A VHDL module in its online variant (left) and the web based simulator interface (right)

The learning process is divided into several lessons, which usually consist of an introduction, several learning steps dealing with the topic, conclusions and a training part. For the VHDL training sessions it is preferred to use the online variant of the modules to achieve a comfortable integration of the simulator, which is only a mouse click away and provides the appropriate partial source code, if there is one, which has to be completed. The other output formats can be used for offline training or as a reference.

All necessary tasks for development and testing of VHDL source codes are done by using our web based simulator interface. Figure 1 shows the appearance of the interface after successful compilation and simulation. Source code is transferred to the server either as locally edited file or as typed (or pasted) input in text fields on the simulator pages. For both methods it is often the case that parts of the VHDL code for an exercise are given and have to be completed with correct code lines to solve the task. After receipt of the source codes by the server, they are preprocessed by a script and compiled and simulated by two separate executables, which are part of the freely available command line tools of Symphony EDA [10]. The console outputs of the compilation and simulation processes are returned in text fields on the resulting web page, which also contains a Java applet for interactive graphical visualization of the simulation results. In case of compilation errors a list of all error messages is presented.

Additionally to this uniformity of development it is possible to track the actions of students on the server to follow their advancement process. The same is possible with the online variants of the modules, since they are hosted on our own module delivery platform.

The <ML>³ features to implement different intensities and targets (student, teacher) have been used to select, which of the tasks are presented and if the solutions of the tasks are included in the practice modules (for the teacher). During winter term 2003/2004 these modules together with the corresponding theory modules and simulation platform have been used in class in order to evaluate them for their application in courses to be held at many different universities in Germany.

The material included in the practice modules is an extension of the Hypermedia VHDL Learning System [4] and represents another step in our long lasting experience in the field of modern ways of education in the area of VHDL synthesis.

4.   CONCLUDING REMARKS

The ability of individual adjustment of learning objects is an important factor for their quality and re-usability. The XML-based document description language <ML>³ implements this concept by utilization of a three-dimensional model and additional didactic exchangeability. Their benefits have been proven in the development of a <ML>³-based course on VHDL, which showed significant improvements compared to former solutions. As a conclusion, one might say: We successfully got in touch with XML, and now our students are able to successfully get in touch with VHDL.

REFERENCES

[1]   K. Noelting, D. Tavangarian, "New Learning Scenarios - Mobile Learning and Teaching at Universities", Proceedings of e-Learn 2003.

[2]   U. Lucke, D. Tavangarian, "Turning a Current Trend into a Valuable Instrument: Multidimensional Educational Multimedia Based on XML", Proceedings of Ed-Media 2002

[3]   "Multidimensional LearningObjects and Modular Lectures Markup Language", http://www.ml-3.org/

[4]   H. Dicken, M. Koch, D. Tavangarian, "The Hypermedia VHDL Learning System - Description and First Experiences", Proceedings of EWME 1996

[5]   „VHDL-Online", http://www.vhdl-online.de/

[6]   "Desktop VHDL", Synopsis, http://www.synopsys.com/services/education/online_learning_tr.html

[7]   "Online Knowledge Center", Mentor Graphics, http://www.mentor.com/es/elearning/

[8]   Esperan Masterclass , "The Multimedia HDL Tutorial", http://www.esperan.com/mc_de.html

[9]   W. Kalfa, R. Kroeger, F. Koehler, "Integrating Simulators and Real-Life Experiments into an XML-based Teaching and Learning Platform", Proceedings of Ed-Media 2002

[10]   Symphony EDA, http://www.symphonyeda.com/

A TIME AND COST- EFFECTIVE MULTIPURPOSE LAB ENVIRONMENT FOR DIGITAL DESIGN COURSES

YI K., CHO Y-S., HAN Y-S.
School of Computer Science and Electronic Engineering, Handong Global University, Pohang, Gyungbuk, 791-708, Korea

1. ABSTRACT

This paper presents a multipurpose and low-cost laboratory environment for digital design courses with emphasis on the concept of reusable design. Suggested design environment for laboratory consists of a low-cost hardware platform, software tools, and design libraries. Using the proposed design environment it is possible to provide all levels of digital design lab courses from the introductory logic design lab for sophomores to the complex system design courses for the seniors. It was also possible to demonstrate the economical and technical efficiency in operating the classes of digital system design using our proposed design environment.

2. WHY A NEW LAB ENVIRONMENT FOR DIGITAL DESIGN?

Modern education of digital system design is under the pressure of changes by the following factors:
- HDL-based design,
- the higher level of complexity and integration,
- (pervasive) use of computers in all aspects of design,
- emerging reconfigurable devices such as FPGAs,
- IP-based design,
- SOC (System-On-a-Chip) design environments, and
- complex and various input/output interface types.

In the past, our classes were focused on the operational principles of microelectronic devices and circuits in the digital microelectronics education. In recent years, the rapid change of design environments and the market needs for the higher level of design complexity and integration require design engineers to have the perspectives of system-level design. Besides, the emerging technology of the implementation devices and the enhancement of design automation tools made it possible for hardware designers to concentrate on the system level design. Thus, the education focus should be moved to equipping the students with the concepts and experiences of the system level design rather than spending most of their time to learn wiring and testing circuit networks on a breadboard.

Here, we present an integrated and well-prepared design lab environment to meet these needs in digital system design classes. Our proposed system consists

A.M. Ionescu et al. (eds.), Microelectronics Education, 107–111.
© 2004 *Kluwer Academic Publishers.*

of hardware platform with FPGA, software suits to interact with FPGA on prototype board, and design libraries to provide complex input/output interfaces as
shown in Figure 1.
The strengths of our design education environment are:
  - low-cost system by adopting(developing) a low-cost and reusable
    hardware platform,
  - multipurpose lab environment that covers from the sophomore labs to
    the senior projects,
  - time-effectiveness in building any digital system using the complex
    input/output interfaces in the forms of libraries
  - time saving in learning design environments because of its simplicity of
    the board configuration and the assisting tools, and
  - intuitive layout of the rich set of input and output interfaces on the
    board.

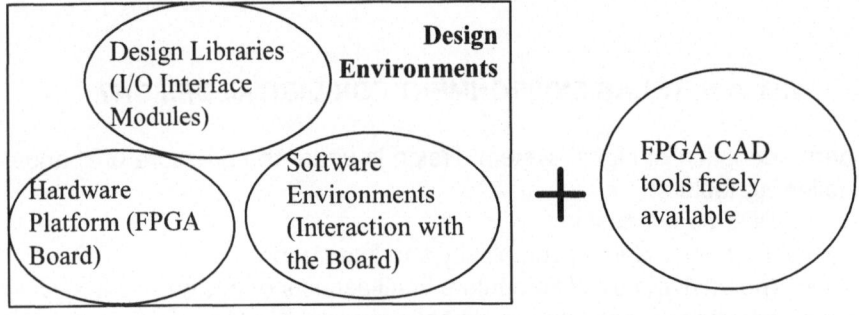

Fig. 1: The Low-Cost Laboratory Design Environments with free FPGA CAD tools

3.   HARDWAER PLATFORM: LOW COST FPGA PROTYPING BOARD

FPGA chips are being used extensively because of the drastic increase in logic
capacity and price drop of the FPGA chips together with its varieties of the embedded functions such as embedded memory and interface logic. Also its reus-
ability, easiness in configuration and availability of the free CAD tools from the
FPGA vendors enabled the students to build their own systems in the class laboratory. Therefore, FPGAs and CAD tools became an essential part of the practical
design education and became necessary for the purpose of the reduction of the
laboratory expenses.
There are several FPGA prototyping boards for digital system design courses
[1,2]. But, some are so simple that they may require additional I/O interfaces, and
it makes the students spend their time and energy on building the electronic cir-
cuitries. The others are too expensive and too complicated for undergraduate

courses. Such complex board organization as DRAMs, PCI interfaces, and dynamic real time debugging features let the students spend too much time on understanding the usage of the boards themselves.

Our rapid prototyping board has the feature of simple and low-cost organization with enough capability to meet the specifications for the various applications. It is ready for future expansions that may be required for undergraduate courses. This board is designed to mount a low-cost Xilinx FPGA device with up to 200K logic capacity. Figure 2 shows the FPGA board we have developed. The board contains two external SRAMs as well as I/O circuits for users: 32 LED arrays, 7-segment displays, 12 input buttons, 4*4 key matrix, a buzzer, an LCD display, and various ports like serial, parallel, PS/2 and VGA. Thus, it is possible for students to perform the experiments of the various levels of digital hardware design in a classroom.

The eight 7-segment displays and 4x4 keypad matrix are organized in a dynamic refreshing and scanning fashion to reduce the use of the limited number of FPGA I/O pins. We chose the package types that achieve lower costs while providing many types of I/O interfaces with dynamic connection.

The board also provides many options for FPGA configuration: configuration using serial PROM on board, configuration using iMPACT (or Hardware Debugger) from Xilinx through parallel DLC-5 cable, or configuration method with our own software with normal parallel cable.

Fig. 2: Developed FPGA Rapid Prototyping Board

4.   SOFTWARE ENVIRONMENT: INTERACTION TOOLS WITH BOARD

We have provided two software tools to interact with the FPGA board: one for configuration tool and the other for running the board with injected stimuli and reading the output signals. Configuration software is provided to download the bit-stream file for FPGA configuration onto FPGA chip and the contents of SRAMs on board via parallel port. It is also possible to upload the contents of SRAMs on the board and to save them as a file using the tool.

A verification tool is also provided with the features of stimulus injection and logic analyzer functions. The tool sends test vectors and receives the result signals through RS-232C cable connected to the board. The circuits under test are wrapped by a virtual wrapper, compiled and downloaded together with wrapper. Using this verification tool set, it is possible to test the designed digital system with larger number of inputs/outputs like 128-bit cryptography processor with more than 256 ports. The developed verification techniques are so efficient that it shows reasonable operation speed and lower hardware overheads compared with IEEE-JTAG standard method.

5.   DESIGN LIBRARIES: I/O INTERFACE MODULES

The board adopts various types of I/O interfaces such as PS2 keyboard, mouse, VGA display, text LCD display, buzzer, Key pad matrix with dynamic scanning feature, serial port, dynamic refreshing 7 segment arrays, and parallel ports. We devised a set of library modules for the complex input/output interfaces to relieve students of the burden to design interface modules at the sophomore level classes. So, students can show their work with various types of outputs like VGA monitor or text LCD, and inputs like PS2 keyboard and mouse, serial communication link from computer or other FPGA board.

6.   APPLICATION TO THE CLASSES

Following list shows the courses that adopt the proposed design environment in the class and topics that were covered at our school. It demonstrates the possibility of our proposed design environment for broader educational applications.

- Logic design lab (sophomore): calculator, digital clock, dice game, elevator controller, slot machine, piano synthesizer, digital lock, and etc.
- Computer design (junior): 8-bit microprocessor, 32/16-bit MIPS-like RISC processor core as in [3].
- Digital system design and individual study (senior): game machine with PS/2 keyboard and PS/2 mouse interface logic, VGA controller, 8-bit UART
- Individual study (graduate level): 128-bit block cipher processor engine

The two external SRAMs of our board can be used either as two separate 8-bit memories for data and instruction to emulate the pipelined Harvard architecture as shown in Figure 3 or as a single 16-bit memory to emulate the 16-bit Princeton architecture for the computer design class. In the digital system design class, SRAMs can be used as graphic data buffers for VGA application.

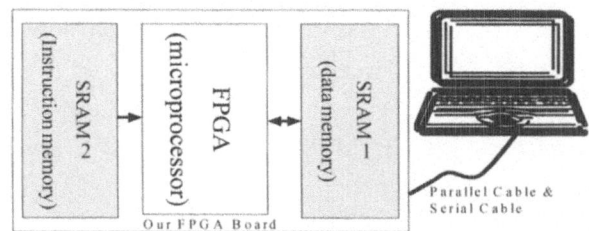

Fig. 3: Configuration of SRAMs for Harvard architecture microprocessor design

CONCLUSION

By introducing the proposed digital design lab environment to the digital design courses we could let the students to spend their time and energy on acquiring the system-level design concept which is essential to the system design engineers. Otherwise, they would have been satisfied with just simulation or would have spent their time in the manual work on the breadboard with TTL chips. We have demonstrated its economical efficiency and capability of our low-cost rapid proto-typing FPGA board and the supporting tools by using them in various courses from sophomore to senior levels.

REFERENCES

[1]   XESS, http://www.xess.com

[2]   E. L. Horta, J. W. Lockwood, et. al, "Dynamic Hardware Plugins in an FPGA with Partial Run-Time Reconfiguration," Proc. of International Symposium on FPGAs, 2002.

[3]   M. Holland, J. Harris, and S. Hauck, "Harnessing FPGAs for Computer Architecture Education," Proc. of International Symposium on FPGAs, 2002.

CMOS AND NANOELECTRONICS

ORAL PRESENTATIONS

TEACHING NANOELECTRONIC DEVICES

FOSSUM J.G.
University of Florida
Gainesville, FL 32611-6130, U.S.A.
(fossum@tec.ufl.edu)

Abstract
As CMOS is being scaled toward the end of the SIA International Technology Roadmap for Semiconductors where gate lengths are projected to be ~10nm, new device structures are, necessarily, emerging. These nonclassical devices include single-gate fully depleted (FD) SOI, double-gate, and triple-gate MOSFETs, all with ultra-thin (<10nm) silicon bodies. And, farther down the road lies the possibility of even more exotic device structures such as silicon nanowires and carbon nanotubes.

This ongoing evolution of nanoelectronic devices is presenting issues regarding the teaching of the devices, as well as their design and fabrication. Strong backgrounds in solid-state physics and quantum mechanics are crucial for good understanding of these devices, but such backgrounds are rare among today's students, as are students genuinely interested in the complex subject. The nonclassical-device teaching issues are overviewed, with references to particular properties of particular devices and particular students, and a new graduate course aimed at dealing with some of them, which is currently being taught, is outlined. The course, having a conventionally scaled classical CMOS device (bulk-Si and partially depleted SOI MOSFETs) design course as a prerequisite, overviews nonclassical nanoscale MOSFETs (FD/SOI and multi-gate devices) that could eventually replace the classical devices as their scaling limit is reached. It includes a computer simulation-based project involving nonclassical devices and/or circuits, in lieu of a final exam, for which physics/process-based compact MOSFET models, linked to Spice3, are made available. Course objectives are to encourage students to work in the device area and, possibly, to produce publishable papers on nonclassical CMOS, as well as to teach the physics-based properties of the novel devices.

A.M. Ionescu et al. (eds.), Microelectronics Education, 117.
© 2004 *Kluwer Academic Publishers.*

THE FUTURE OF CMOS NANOELECTRONICS

DELEONIBUS S.
CEA-LETI/ NANOTEC, sdeleonibus@cea.fr
CEA-Grenoble 17 rue des Martyrs 38054 Grenoble Cedex 09 France

Abstract
Historically, innovations have been possible because of the strong association of devices and materials research. The demand for low voltage, low power and high performance are the great challenges for engineering of sub 50nm gate length CMOS devices. We point out the main issues to address in order to investigate and push the limits of CMOS technology. The alternative architectures allowing to increase devices drivability and reduce power are reviewed. Among the materials options to be integrated, HiK gate dielectric and metal gate are among the most strategic options to consider for power consumption and low supply voltage management. New architectures and options are reviewed through the issues to address in gate/channel and substrate, gate dielectric as well as source and drain engineering. It will be very difficult to compete with CMOS logic because of the low series resistance required to obtain high performance. By introducing new materials, Si based CMOS will be scaled beyond the ITRS as the future System-on-Chip Platform integrating new disruptive devices. Functionality of devices in the range of 5 nm channel length has been demonstrated showing that CMOS technology could still be used in the future if we manage to implement new materials and device architecture options.
Key words: CMOS, Nano, roadmap, HiK, Metal gate, Nanocrystals, Flash memory.

1. INTERNATIONAL TECHNOLOGY ROADMAP OF SEMICONDUCTORS ACCELERATION AND ISSUES.

Since 1994, the International Technology Roadmap for Semiconductor (ITRS)[1] (Figure 1) has been accelerating the scaling of CMOS devices to lower dimensions continuously despite the difficulties that appear in device optimization.
However, uncertainties about lithography, economics and physical limitations can probably slow down the evolution. For the first time, since the introduction of poly gate in CMOS devices process, showstoppers other than lithography appear to be deserved special attention and could require some breakthrough or evolution if we want to continue scaling at the same rate. Design could also be affected by this evolution.

A.M. Ionescu et al. (eds.), Microelectronics Education, 119–132.

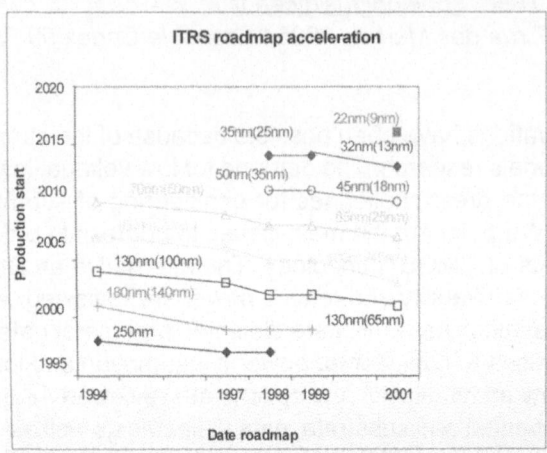

Fig. 1: ITRS roadmap acceleration since 1994[1]. Example for MPU and ASICs

Which are the main showstoppers for CMOS scaling? In this chapter, we focus on the possible solutions and guidelines for research in the next years in order to propose solutions to enhance CMOS performances before we need to skip to alternative devices. In other words, how can we offer a second life to CMOS ?

To that respect, the roadmap distinguishes today three types of products: High Performance(HP), Low Operating Power(LOP) and Low Standby Power(LSTP) devices. In the HP case, a historical fact will happen by the 32nm node: the contribution of static power dissipation will become higher than dynamic power contribution! In this papaer we will analyze the various mechanisms giving rise to leakage current in a MOS device and can impact consumption of final devices. Gate leakage current is already a concern. In the case of LSTP devices, a HiK gate insulator could be needed earlier than expected in order to limit static consumption(see section 4.2).

2.   LIMITATIONS AND SHOW STOPPERS COMING FROM CLASSICAL CMOS SCALING.

Several mechanisms can generate devices leakage in ultra small MOSFETs. which can be sorted in two categories:

a) Classical type
- Drain Induced Barrier Lowering(DIBL) is due to the capacitive coupling between source and drain.
- Short Channel Effect(SCE) due to the charge sharing in the channel in the short channel devices at low Vds.
- Punch-Through between source and drain due to the extension of source space charge to the drain.

b) Quantum and high field effects
- Direct tunneling through the gate dielectric.
- Field assisted tunneling at the drain to channel edge. This effect occurs if electric field is high and tunneling is enhanced through the thinnest part of the barrier.
- Direct tunneling from source to drain. This effect will occur in silicon for a thicker barrier than on SiO2 because the barrier height is lower and the equivalent barrier thickness is higher, due to the higher dielectric constant.

Velocity overshoot and ballistic transport are the mechanisms that will enhance drivability in sub 100nm channel lengths devices. However, the impact of scattering by dopants on transport is not negligible even in 5nm range channel lengths [2][3]. Superhalo is efficient to improve SCE and DIBL in 16nm finished gate length(Figure 2)[4] This effect is balanced by the dopant diffusion effect on the channel transport properties degradation.

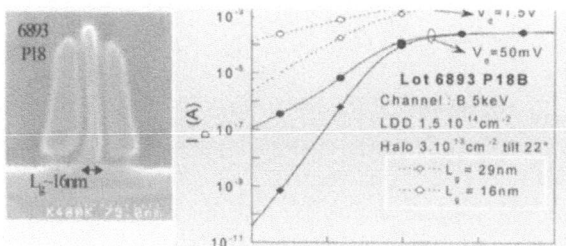

Fig. 2: Functional Finished gate length 16nm MOSFET sub threshold characteristics. Gate oxide thickness is 1.2nm[3]. Isat is 600μA/μm

3.   ISSUES IN LOWERING SUPPLY VOLTAGE.

In the future, the electronics market will request portable objects used in daily life and thus low standby and active power dissipation. For sub 0.10μm devices, the following main issues cannot be avoided:

3.1   *DIRECT TUNNELING THROUGH SIO2*

is significant for a thickness less than 2.5 nm. It contributes to the leakage component of power consumption. 1.4 nm thin SiO2 is usable without affecting devices reliability[5][6],[7],[2].

3.2   *HIGH DOPING LEVELS IN THE CHANNEL*

reaching more than 5×10^{18} cm^{-3} enhances Fowler-Nordheim field assisted tunneling reverse current in sources and drains up to values of 1A/cm2 (under 1V)[8].

3.3   CLASSICAL SMALL DIMENSION EFFECTS

are more severe than the fundamental limits of switching (quantum fluctuations, energy equipartition, or thermal fluctuations). A minimum value is required for threshold voltage due to:

* *subthreshold inversion.* For ideal fully-depleted SOI(FDSOI) 59.87 mV/dec subthreshold swing can be obtained at 300K. The limit VT value is 180mV precluding a supply voltage V_S lower than 0.50V. Impact Ionization MOS (I-MOS) would allow to reduce subthreshold swing to 5mV/dec. However, performance remains an issue[9].

* *short channel effect* due to the charge sharing along the transistor channel following the relation:

$$\frac{C_w}{C_{ox}}\frac{x_j}{L}\left[\left(1+2\frac{W}{x_j}\right)^{1/2}-1\right]=-4\varphi_F\frac{\varepsilon}{\varepsilon_{ox}}\frac{t_{ox}}{L}\frac{x_j}{W}\left[\left(1+2\frac{W}{x_j}\right)^{1/2}-1\right] (1)$$

Here VT is expressed by: $V_T=V_{FB}+2\varphi_F-\dfrac{Q_B}{C_{ox}}$ (2)

where $V_{FB}=\varphi_{MS}-\dfrac{Q_{ox}}{C_{ox}}$ (3) and $C_{ox}=\dfrac{\varepsilon_{ox}}{t_{ox}}$;

$$\varphi_{MS}=\varphi_M-\varphi_s \ (3.1)$$

ΔV_T is the threshold voltage decay; t_{ox} is the gate dielectric thickness; ϵ and ϵ_{ox} are the silicon and gate dielectric constant respectively; L is the channel length; X_j is the drain or source junction depth; W is the space charge region depth; V_T is the threshold voltage ; V_{FB} the flatband voltage ; ϕ_F the distance from Fermi level to the intrinsic Fermi level ; Q_B the gate controlled charge ; C_{ox} is the unit area capacitance of the gate insulator. φ_{ms} is the difference between the extraction potentials of the gate and the semiconductor; Q_{ox} is the oxide charge density ; φ_M and φ_S are the metal and the semiconductor workfunction.

Gate depletion and quantum confinement in the inversion layer will play an important role on short channel effect by adding their contribution to the gate to channel capacitance C_G. SCE is the main limitation to minimal design rule. For low VT values it can be of the order of VT. In order to maintain inverter delay degradation

to less than 30% , we must observe the condition $V_T=\dfrac{V_S}{3}$ [10].

3.4   STATISTICAL DOPANT FLUCTUATIONS.

The effect of dopant fluctuations has already been considered by Schockley in 1961[11]. Recently, special attention is being paid to this subject because the number of dopants in the channel of a MOSFET tends to decrease with scaling[12], [13]. The random placement of dopants in the MOSFETs channel by ion implantation will affect devices characteristics for geometries lower than 50 nm. The discrete nature of dopant distribution can give rise to devices characteristics asymmetry[13].

Dopant fluctuations and Fowler Nordheim limitation at high electric field will encourage the use of low doped thin SOI.

4.   MOSFET OPTIMIZATION

Possible solutions to overcome the physical limitations encountered in classical scaling are reviewed through:
- - gate stack and channel/substrate engineering
- - source and drain engineering
- - gate dielectric engineering

4.1   GATE STACK AND CHANNEL /SUBSTRATE ENGINEERING.

4.1.1 Gate and channel. Issues in classical scaling of bulk MOSFET

Gate and channel engineering must be optimized together because both physical characteristics affect the nominal VT value of expression (2) which can be written as :

$$V_T = V_{FB} + 2\Phi_F - \frac{Q_B}{C_G} \ (4)$$

(gate depletion and channel quantum effects are taken into account).
Low VT values will result from:

4.1.2 adjusting gate insulator thickness(see section 4.2).

4.1.3 tuning surface doping concentration

as low as possible. Excellent localization of the dopant profile is needed to minimize junction parasitic capacitance and body effect. Selective Si epitaxy of the channel has also been suggested to achieve almost ideal retrograde profiles[14].

4.1.4 Strained channel engineering

Strained SiGe[15], SiGexCy based alloys or strained Si epitaxy have been studied to increase the channel mobility [16][17]. However, high quality gate insulator and subthreshold characteristics optimization require a Si cap layer on top of the channel and low thermal budget[17]. HiK gate insulator is needed in these architectures[18]. Selective epitaxial Si:C acts as Boron diffusion barrier (Figure 3) and thus help to improve drastically short channel effect[17] as well as low field mobility.

S. DELEONIBUS

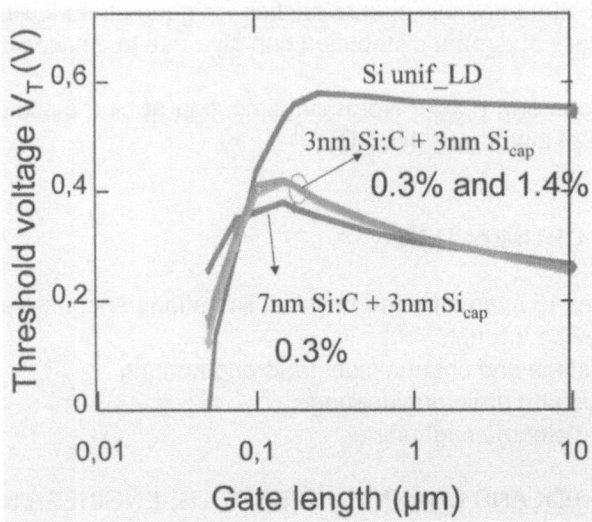

Fig. 3: Effect of introduction of Carbonated silicon in MOSFET channel on Short
Channel effect.[17]

4.1.5 choosing the gate material.

Ideal transfer CMOS inverters characteristics requires symmetry of threshold voltage for n and p channel devices (i.e.VTP=-VTN). Several alternatives have been
envisaged :

- *the use of n+ poly gate for nMOSFET and p+ poly gate for pMOSFET.*
  This solution suffers from Boron penetration into SiO2 coming from the
  p+ doped gate. Nitrided SiO2 *limits without avoiding* this effect:
  trapping centers are created near or at the SiO2/Si interface
  decreasing carrier mobility.
- *the use of metal gate material.* No gate depletion is observed in this
  case. The use of midgap gate(TiN for example)on bulk or partially
  depleted SOI will be dedicated to supply voltages higher than 1V.
  Workfunction engineering for Dual metal gates, is challenging: the
  highest CMOS performance/lowest leakage current trade off can be
  obtained. It is mandatory on low doped FDSOI.

Several approaches have been proposed for metal gate integration. The classical
process integration requires the protection of the metal gate material from ion im-
plant as well as oxidation during the dopant activation anneal. TiN is often chosen
as a gate material[19], because it is available as a standard in the industry. Integration with Ta2O5[20,21]has been reported : however leakage current is an important issue. Alternatives such as the damascene gate (Figure 4)[22,23] have
been achieved in order to avoid the source and drain activation temperature

issue. High Frequency and Multi threshold devices could be embedded in Systems On Chip thanks to the damascene architecture.

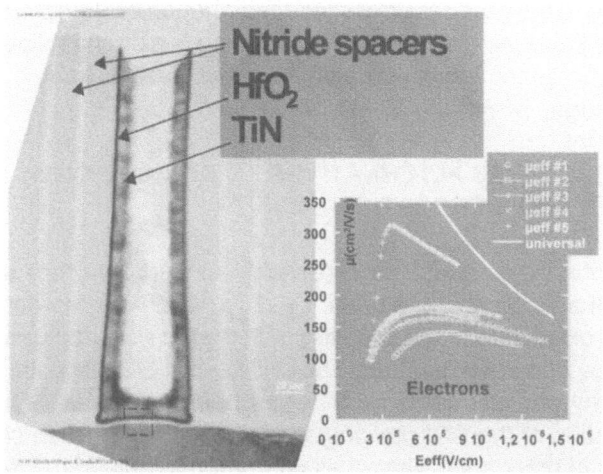

Fig. 4: TEM cross section of TiN/HfO2Damascene gate stack[23]. Electron Mobility degradation with HiK inserted

4.2 GATE DIELECTRIC ENGINEERING.

The gate leakage due to direct tunneling in standard SiO2 or SiOxNy is one major show stoppers[1]. It will impact directly the static power dissipation Pstat according to the relation :$P = P_{stat} + P_{dyn}$ (5)$P_{stat} = V_{dd}xI_{off}$ and $P_{dyn} = CV_{dd}^2 f$ (6)

P being the total power dissipation; P_{stat} being the static power dissipation; P_{dyn} being the dynamic power dissipation. If one considers a circuit with active area of the order of 1cm2 and gate oxide tox=1.2nm . If Ioff is due to gate leakage, then considering Vdd=0.5V then Pstat(0.5V)= 5W. We would get Pstat(1.5V) = 750W for a Vdd of 1.5V!! This is a major show stopper for scaling of CMOS technology. That is why High K will be needed in the near future. Besides affecting static power, gate leakage impacts also negatively delay time [24]and affects the functionality of logic circuits.

A decrease of devices performance has been reported if SiO2 thickness is lower than 1.3nm[25] suggesting a surface roughness limited mobility process due to the proximity of sub-oxide. The strong band bending due to quantum mechanical corrections affects the lower limit of supply voltage in the constant field scaling approach[26]. Solutions compatible with silicon gate are also investigated to keep compatibility with a standard CMOS process flow: HfSiOx, ZrSiOx are given much attention as good candidates[27]. These solutions are *dielectric thickness budget* consuming(SiOx interface) and Fermi level pinning occurs at the HiK/poly gate interface[28].

Recently, the lowest leakage current has been reported by using 1.3nm EOT HfO2 combined with a TiN gate integrated on 45 nm CMOS by a damascene process[23] (figure 4). Electron mobility degradation is reported compared to SiO2 gate dielectric[23] attributed to stress induced phonon scattering. These materials have a smaller bandgap than SiO2: thus trapping is a strong reliability issue. That is why a SiON interface could be helpful to reduce the leakage current thanks to the higher bandgap of SiON.

4.3 ARCHITECTURES ALTERNATIVES TO IMPROVE CMOS PERFORMANCES AND INTEGRATION DENSITY.

In order to obtain the lowest subthreshold slope(60mv/dec) and acceptable DIBLon FDSOI a practical rule is used: $T_{Si} \leq L_{gate}/4$[29]. The spreading of potential into the buried oxide, due to the coupling with the top gate, increases the coupling between source and drain and thus DIBL. Ultra-thin SOI films are difficult to control. That is why partially depleted SOI has been proposed[29,30]. Because of complete isolation of the SOI devices as well as lower junction capacitance, improved figures of merit are obtained as compared to bulk[29]. The threshold voltage is dependent on Si film thickness whenever the film thickness becomes lower than the space charge region. VT is expressed as[29]:

$$V_T = V_{FB} + 2\varphi_F + \frac{qN_A T_{Si}}{2 C_{ox}} \quad (7)$$

N_A is the acceptor concentration; T_{Si} is the silicon thickness; C_{ox} is the gate insulator capacitance.

Scaling of FD devices encounters some limitations due to the quantum confinement in ultra thin films and its incidence on the threshold voltage value[31]: the increase of the fundamental level of the conduction band will increase flat band voltage and VT consequently.

Recently, the functionality of ultra small 6nm gate length devices on 7nm thin Si film was demonstrated[32].

Self-heating is an issue on fully isolated devices because of bad SiO2 thermal conductivity. Replacing SiO2 by Al2O3 has been proposed as a solution because the thermal conductivity of Al2O3 is ten times larger than for SiO2[33],[34]

SOI material should allow to realize attractive devices like multi gated MOSFETs[35] that will allow further scaling of FD depleted devices which are limited by the quantum confinement issue and DIBL via the coupling of the gate with buried oxide [31]. With multi gate devices, short channel effects and leakage current can be drastically reduced because 60mV/dec subthreshold swing and high drivability can be obtained. Transport occurs by volume inversion due to the coupling of both gates. The conditions to control short channel can be relaxed compared to single gate FD devices[31][36], [37],[38],[39],[40]. Nevertheless, the control of thin SOI and design of high density circuits with these devices have to be demon-strated.

The main feature of these devices is to bring a solution to the channel dopant fluc-tuation problem. Reducing the film thickness to the minimum, allows to use nearly intrinsic Si films because bulk punch-through is no more a problem. Adjusting VT to match overdrive with a low supply voltage will require to adjust the gate work-function φ_M according to relation (3.1). That is why, workfunction engineering on metal gate and HiK stacks is mandatory for low VS applications.

4.4   SOURCE AND DRAIN ENGINEERING.

Low energy (<1keV)[25] and heavy molecules (BF3[41], B10H14[42],...)are the easiest ways to replace Boron to achieve p+ shallow junctions. Plasma doping is investigated as an alternatives to obtain lower than 25nm as implanted p+ junc-tion depths[43]. Transient Enhanced Diffusion (TED) is still limiting process to reach the specified final junction depths. Fast ramp up and down -so called spike annealing- must be combined with Low Energy Ion Implantation [43] to reduce TED as much as possible, by reducing the role played by extended and dopant defects. Excimer Laser Anneal[44],[45] has demonstrated the best trade off be-tween low sheet resistance and junction depth shallowness: highest solid solubil-ity combined with fast processing can be achieved. Low sheet resistance combined with low silicon consumption can be obtained with monosilicides(Ni-Si,PtSi,) instead of disilicides (TiSi2, CoSi2)[46]. Devices on thin SOI will require raised sources and drains to facilitate silicidation.

5.   ALTERNATIVE CMOS OR ALTERNATIVE TO CMOS?

Many research teams are making efforts on Single Electron Transistors(SET)op-eration based on the Coulomb blockade principle. Demonstration of CMOS in-verter operation at 27K has been achieved by using a Vertical Pattern Dependent Oxidation (V-PADOX) process[47]. No solution has been found that could com-pete with CMOS devices. Some possibilities to achieve memory functional devic-es by using single electron trapping by a Coulomb blockade effect for DRAM [48], or Non Volatile applications[49], [50], [51]. have been pointed out.

This effect supposes that the Coulomb energy $\dfrac{e^2}{2C}$ (8)

is larger than the thermal energy of electrons kT(e is the electron charge; C is the capacitance of the quantum box). This energy is necessary to localize electrons in a Coulomb box provided that tunneling is the limiting process: implicitely, one has to use very low capacitance and sufficiently high tunneling resistance. How-ever, the Coulomb blockade process will be self limiting by charge repulsion which reduces the speed of the charge transfer. Non Volatile Memory applications can be envisaged by using trapping in nanometer size Si quantum dot[50]: Al2O3 has been chosen as the tunnel insulator with reasonable interface states density(less than 1011 cm-2) and can also increase the dot density as compared to other materials(in the range of 1012cm-2).

Whether the involved writing or erase mechanisms are due or not to single electron transfer has been a controversial debate. If the Si dots are randomly distributed in large area devices then it is very difficult to identify whether the single electron transfer is occurring or not due to the large number of dots. It is thus very important to use a device of the smallest size possible to get a high sensitivity to single electron transfer in one dot or a low number of dots. Such a result has been obtained at room temperature on 20nmx20nm Non Volatile Memory Silicon wire based on Silicon quantum dots(Figure 5b)[51]: current spikes on the writing or erasing characteristics have been identified as single electron trapping or detrapping respectively.

Coulomb blockade oscillations can be observed if the series access resistance with the quantum well is high enough compared to the

resistance quantum $(\dfrac{e^2}{h})^{-1}$ [52] (9).

This effect has already been reported on 50nm gate length N channel MOS transistors at 4.2K[53] making CMOS transistors attractive as single electron devices candidates. As gate length is scaled down to 20nm, access resistance becomes larger and channel conductance oscillations appear at higher temperatures(here 75K) (figure 5b) [4].

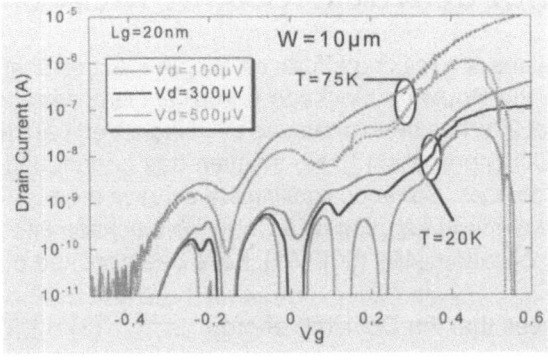

(a)

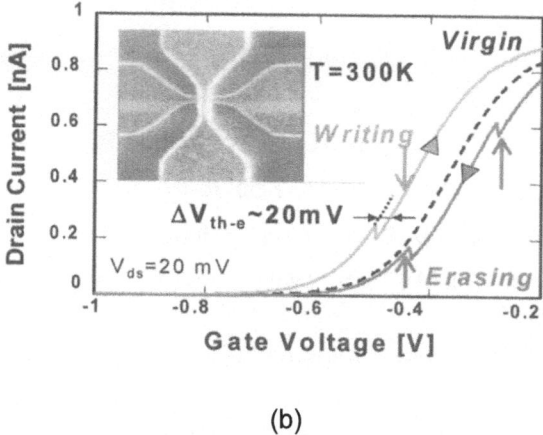

(b)

Fig. 5: Devices characteristics evidencing Single Electron phenomena (a) Drain current oscillations in a Lg=20nm MOSFET at 75 and 20K, demonstrating that Coulomb blockade is possible in such devices[3] (b) Writing and erase characteristics of 20nmx20nm(WxL) devices at room temperature. Spike in Id(Vg) characteristics is due to trapping or de-trapping of one electron in SOI nanowire Si dot Memory. Top view of 20nmx20nm nanowire[51] inserted.

CONCLUSIONS

Beyond the roadmap, multigate devices using strained channels will be widely used for high performance CMOS. Si based alloys or compatible semiconductors will be introduced to enhance the possibilities of future Systems on Chip. Single electronics will be a major study subject to optimize the use of ultra small devices.

REFERENCES

[1]   The International Technology Roadmap for Semiconductors (2001 edition)

[2]   S. Deleonibus et al. IEEE Electron Dev. Letters, pp173- 175, April 2000.

[3]   G.Bertrand et al., Silicon Nanoelectronics Workshop 2000, pp10-11, June 10-11, 2000, Honolulu(HI)

[4]   G.Bertrand et al., 4th Workshop on Ultimate Integration of Silicon Proceedings, pp 10-13,   March 2003, Udine(Italy).

[5]   S.Deleonibus et al., ESSDERC Tech. Digest 1999, pp.119-126,Leuven, Sept. 1999

[6]   H.Iwai et al IEDM Tech Digest 1998, pp.163-166, Dec. 1998, San Francisco(CA)

[7]   C.Caillat et al VLSI Technology Symposium Tech. Digest 1999, pp. 89-90, June 1999, Kyoto(Japan).

[8] Y.Taur et al IBM Journal of Research and Development, vol. 39,N1/2,pp.245-260, 1995.

[9] K.Gopalakrishnan et al. IEDM2002 Tech. Digest, pp.289-291, Dec. 2002, San Francisco(CA).

[10] T.Oyamatsu et al. Tech Digest VLSI Symp. pp.89-90, June 1993

[11] W.Schockley, Solid State Electron.,2,pp.35-67, 1961

[12] V.De et al Tech Digest VLSI Symp. pp. , June 1998

[13] W.Wong et al IEDM Tech Digest 1993, pp.705-708, Dec. 1993, Washington(DC)

[14] T.Ohguro et al., IEEE Trans. Electron Dev. , vol. 45,n°3, pp. 710-716, March 1998

[15] M.Carroll et al., IEDM Tech Digest 2000, pp.145-148, Dec. 2000, San Francisco(CA)

[16] K.Rim et al., IEDM Tech Digest 1998, pp.707-710, Dec. 1998, San Francisco(CA)

[17] T.Ernst et al., VLSI Technology Symposium 2003 Tech. Digest, pp 51-52, Kyoto(Japan)

[18] K.Rim et al., VLSI Tech Symp, Honolulu, June 2002

[19] J.Lee et al. , Tech Digest VLSI Symp.,p. 208, June 1996

[20] A.Chatterjee et al., IEDM Tech Digest 1998, pp.777-780, Dec. 1998, San Francisco(CA)

[21] [T.Devoivre et al., Tech Digest VLSI Symp,p131,June 1999

[22] A.Yagashita et al. IEDM Tech Digest 1998, pp.785-788, Dec. 1998, San Francisco(CA)

[23] B.Guillaumot et al., IEDM 2002 Tech. Digest, pp. 335-338, Dec 2002, San Francisco(CA)

[24] D.Souil et al 3rd ULIS Workshop 2002, pp139-142,March 2002, Munich(FRG)

[25] G.Timp et al, IEDM Tech Digest 1998, pp.615-618, Dec. 1998, San Francisco(CA)

[26] S.Takagi et al., IEDM Tech Digest 1998, pp.619-622, Dec. 1998, San Francisco(CA)

[27] J.Lee et al. IEDM Tech Digest 1999, pp.133-136, Dec. 1999, Washington(DC)

[28] C.Hobbs et al., VLSI Technology Symposium 2003 Tech. Digest, pp 9-10, Kyoto(Japan)

[29] JL Pelloie, ISSCC Tech Digest 1999, p.428, San Francisco(CA) Feb. 1999

[30] L.Leobandung et al. IEDM Tech Digest 1998, pp.403-407, Dec. 1998, San Francisco(CA)

[31] Lolivier et al ECS Spring 2003, April 2003, Paris(France)

[32] B. Doris et al., IEDM 2002 Tech Digest, pp267-270, Dec 2002, San Francisco(CA)

[33] H.Nakayama et al., IEEE SOI conf.2000 Proc, pp.128-129, Oct 2000, Wakefield(Mass)

[34] K.Oshima et al. SOI Conference 2002 Tech Digest, pp 95-96, Oct 2002,

[35] H.S.P. Wong et al., ,IEDM Tech Digest 1997, pp.427-430, Dec. 1997, Washington(DC)

[36] F.Allibert et al., ESSDERC 2001, Sept 2001, Nurnberg(FRG)

[37] B.Yu et al., IEDM 2002 Tech Digest, pp 251-253, Dec 2002, San Francisco(CA)

[38] J. Kedzierski et al., IEDM 2002 Tech Digest, pp247-250, Dec 2002, San Francisco(CA)

[39] FLYang et al., IEDM 2002 Tech Digest, pp255-258, Dec 2002, San Francisco(CA)

[40] R.KWGuarini et al., IEDM 2001 Tech Digest, pp 425-428, Dec 2001, Washington(DC)

[41] J.M.Ha et al. ,IEDM Tech Digest 1998, pp.639-642, Dec. 1998, San Francisco(CA)

[42] K.Goto et al. ,IEDM Tech Digest 1997, pp.471-474, Dec. 1997, Washington(DC)

[43] M.Takase et al. ,IEDM Tech Digest 1997, pp.475-478, Dec. 1997, Washington(DC)

[44] D.F.Downey et al., Proc. Of Mat. Res. Soc.,525 , 263(1998)

[45] T.Noguchi et al., Proc. Of Mat. Res. Soc., 146, 35(1985)

[46] C.Laviron et al., Ext. Abst. 2nd IWJT, IEEE-Cat.No.01EX541C, 2001: 91-4., Nov 2001, Tokyo(Japan)

[47] T.Ohguro, ECS Symp on ULSI 1997, p275, Oct. 1997, Montreal(CA)

[48] Y.Ono et al. , ., IEDM 2001 Tech Digest, pp 367-370, Dec 2001, Washington(DC)

[49] S.Tiwari et al., Appl. Phys. Lett. 68, 1377(1996)

[50] K.Yano IEDM Tech Digest 1998, pp.107-110, Dec. 1998, San Francisco(CA)

[51] A. Fernandes et al., IEDM 2001 Tech Digest, pp 155-158, Dec 2001, Washington(DC)

[52] G.Molas et al. WODIM 2002 , Nov 2002, Grenoble (France)

[53] M.Sanquer et al. SNW 2003, pp70-71 Kyoto (Japan)

[54] M.Specht et al. IEDM Tech Digest 1999, pp. 383-341 , Dec. 1999, Washington(DC)

CMOS ANALOG CIRCUITS DESIGN EDUCATIONAL TOOL

KAYAL M., STEFANOVIC D., PASTRE M.
Electronics laboratories, Ecole Polytechnique Fédérale de Lausanne, STI, IMM –
LEG, CH - 1015 Ecublens, Switzerland, maher.kayal@epfl.ch

Abstract:
This paper presents a standalone educational tool dedicated to the understanding of the MOS transistor behavior from electronic circuits design point of view. It encapsulates a chart-based design environment aimed at the design of analog circuits starting from basic analog structures (one transistor or groups of transistors) and going towards basic analog cells such as: OTA, operational amplifiers, voltage regulators.

1. introduction

With the evolution of CMOS technology, the behavior of the MOS transistor has become more and more complicated. However, the analog circuits design methodologies presented in the books, lecture notes or teaching references are still based on simplified interpretations. The proposed hand calculation methods are efficient only in some special cases. As a result, the analog designers use Spice simulations with the complicated mathematical models and rely on their own expertise and intuition. Moreover, there are no good hand calculation methods available for MOS transistor. At the same time, the teaching of analog circuit design requires a good understanding of the transistor behavior as a starting point.

The proposed tool presents a new interactive knowledge-based design methodology for analog structures sizing. It assists the student by presenting the necessary theoretical knowledge and the tips from an experienced designer. Furthermore, its interactive interface provides instantaneous visualization of the design tradeoffs. A transistor level calculator is capable of exploring complex relations and displays the results on charts, which the user can interact with. In this way, this tool shows all the dependencies and characteristics of analog structures at one glance and allows the designer to find tradeoffs and optimize the circuits efficiently. The implemented transistor level calculator uses the complete set of equations based on the EKV MOS model [1], which links the equations for weak and strong inversion in a continuous way. Large numbers of the transistor parameters, which are important for analog design, can be extracted from the model: inversion factor, saturation voltage, Spice-like threshold voltage, Early voltage, small signal parameters, parasitic capacitances, g_m/I_D ratio[2]., transconductance efficiency factor.

A.M. Ionescu et al. (eds.), Microelectronics Education, 133–138.

2.   Student environment

Transistor level analog CMOS design leads to the selection of bias current and geometry sizing (W/L) of each transistor in a circuit. The CAD methodology presented here helps the students to understand that a design task consists of exploring transistors' limits and parameters' relations in order to find good tradeoffs. Appropriate graphical representations are used to provide a visual feedback. Furthermore, the user-friendly graphic interface has the same appearance for each basic analog structure. The student can change the current biasing or geometrical dimensions and observe simultaneously the behavior of the other parameters. In this way, he gets an intuitive understanding of the device behavior.

An analog structures library is embedded in this tool, including basic analog structures such as: single transistor, current mirror, differential pair, cascode stage, cascode current mirrors. For each structure a set of general electrical parameters (small signal model, DC biasing values, parasitic capacitances, speed, noise) is calculated and displayed. Some specific parameters (for example maximum DC offset for differential pair, current mismatch for current mirror) are also shown. This enables to analyze the behavior of basic structures in the environment of a given circuit, and observe parameters that are important for design tradeoffs (Fig. 1).

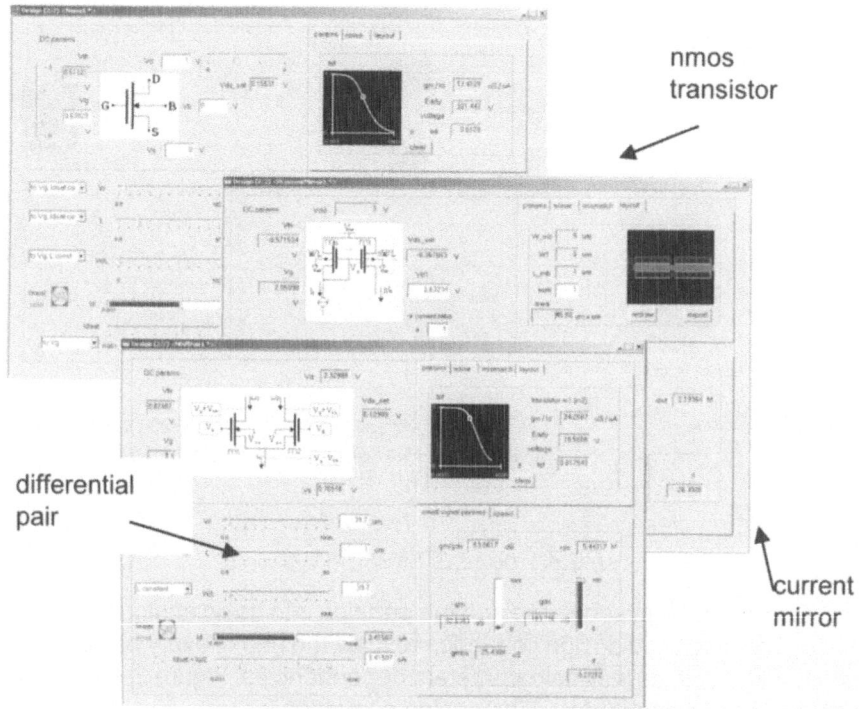

Fig. 1: Basic analog structures transistor level sizing

The procedural design of complex analog structures is also implemented. The current tool version enables systematic design of operational transconductance amplifiers (OTAs) and different operational amplifiers structures. The procedural design flow consists of: topology selection, circuit partitioning into basic analog structures and chart-based analog structures sizing and transistor level design.

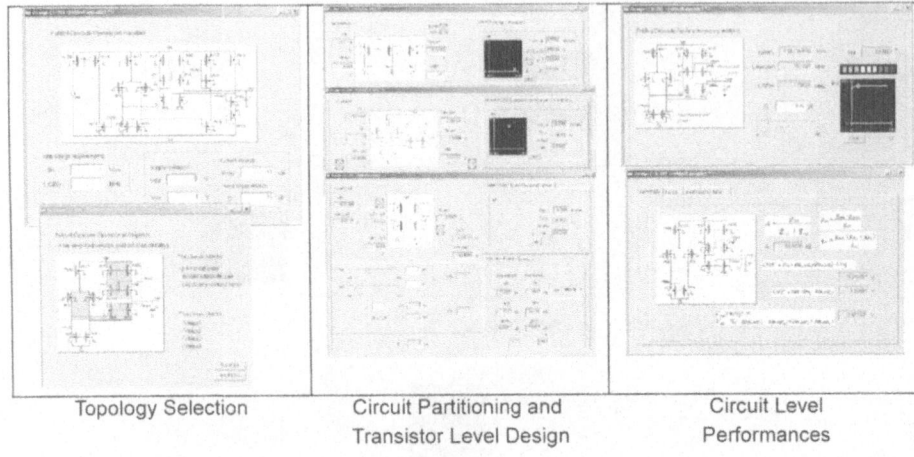

| Topology Selection | Circuit Partitioning and Transistor Level Design | Circuit Level Performances |

Fig. 2: Circuit level design flow

At circuit level, the design sequence and the partition into basic analog blocks are predefined. The proposed design sequence reveals the role of every basic analog structure in the circuit, as well as parameters dependencies. Figure 2. depicts the folded cascode OTA design flow. The design sequence is:

- Initialization with the given technology parameters and topology selection.
- Circuit partitioning.
- Transistor level sizing of the basic analog structures.
- Frequency analysis, circuit summary, noise and offset analysis.

During the design procedure, appropriate graphical views provide immediate and intuitive feedback on circuit performance. Each transistor or related pairs or groups of transistors are identified with the selected drain current, inversion factor and geometrical dimensions and the resulting electrical performances. The final circuit level summary provides a complete documentation of the electrical behavior of the designed cell.

3.   Example of Circuit Design of DC-DC linear voltage regulator

During student lectures, two basic DC-DC voltage linear regulator topologies are introduced: a low dropout regulator (LDO) with a common source output bypass transistor and a non-LDO with a common drain output bypass transistor. The theoretical part emphasizes the cell efficiency as a key problem. In the practical part, the students are asked to design an LDO voltage regulator with the specified maximum output current using a given technology. The tool described above is used to guide the students through this exercise. Figure 3. shows the proposed design steps:

1. The bypass transistor sizing according to DC current capability.
2. The design of the complete error amplifier capable of driving the input impedance of the bypass transistor determined in the previous step. The main goals are the optimization of power consumption and the improvement of PSRR (Power Supply Rejection Ratio). Transistor level design consists of the following steps:

    a) The biasing block design with the effect on slew rate, negative common mode range.

    b) The differential pair design with the effect on gain bandwidth, noise, offset.

    c) The active load design with the effect on noise, offset, gain, stability.
3. The analysis of the stability behavior of the designed LDO voltage regulator cell, taking into account the nature of the load (resistive or capacitive).

At each step, the necessary theoretical guidelines and tips from an experienced designer are presented in the separate window. Furthermore, the student can navigate from one point to the other through the design scenario in order to modify the transistor level design and reach the optimum performance of the designed cell.

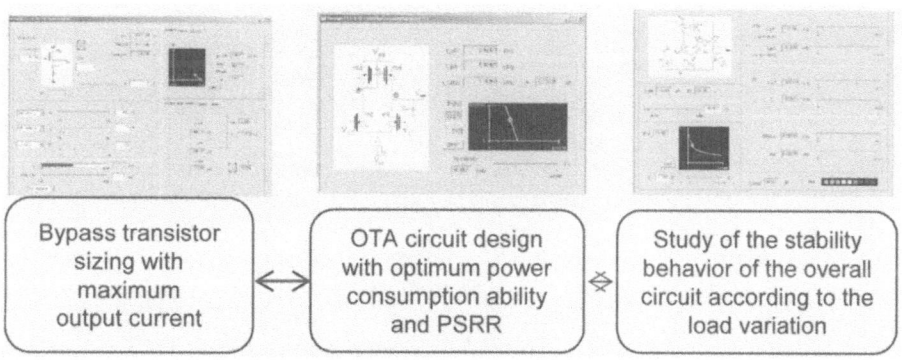

Fig. 3: Low Dropout Voltage Regulator (LDO) design methodology and analysis

Conclusion

This paper presented a new standalone chart-based tool. Its purpose is to encapsulate and hand the design knowledge and experience towards students. This interactive tool enables design and re-design of a wide range of circuits. It is called Procedural Analog Design (**PAD**) and can be downloaded from: **http://legwww.epfl.ch/CSL.**

REFERENCES

[1]   M. Bucher, C. Lallement, C. Enz, F. Théodoloz and F. Krummenacher, "The
      EPFL-EKV MOSFET Model Equations for simulation, Version 2.6"
      Technical Report, EPFL, July 1998, available on-line: http://legwww.epfl.ch/
      EKV

[2]   F. Silveira, D. Flandre, P.G.A. Jespers," A gm/ID based methodology for the
      design of CMOS analog circuits and its application to the synthesis of a
      Silicon-on-Insulator micropower OTA", IEEE Journal of Solid-State Circuits,
      31 (1996) 1314-1319.

PRACTICAL COURSE OF DESIGN, FABRICATION AND TESTING OF CMOS GATE ARRAY

MITA Y., KOMATSU S., IKEDA, M., FUJISHIMA M., ASADA K.
VLSI Design and Education Center (VDEC), the University of Tokyo. Bunkyoku Yayoi 2-11-16, Tokyo 113-8656, Japan http://www.vdec.u-tokyo.ac.jp/

1.   INTRODUCTION

Learning VLSI design is essential for modern engineers. In the Department of Electrical and Electronics Engineering, the Univ. of Tokyo, several courses complementally cover whole of VLSI Design and Fabrication as shown in table 1. In particular, education with physical practice is important and emphasized.

The courses are classified into three categories: lectures in auditorium, seminar with CAD design, and the Gate Array practical course. Students learn frontend design procedure in C-language-based hardware design and FPGA- demonstration seminar [1]. In the VLSI Layout and Simulation seminar, a backend layout design practice is given to students. Students design layout of CMOS logic circuits, perform Design Rule Check, extract SPICE netlist and simulate. Gate Array course covers from backend design to fabrication. Gate Array is the only course where students experience LSI fabrication process.

Table 1: VLSI education courses in the EE department.

Type	Title	Grade	Hour	Area
Lecture	Basics on VLSI Engineering	3rd	21hrs	General
Lecture	VLSI Design Engineering	4th	21hrs	Design
Lecture	Electronics Material Process	4th	21hrs	Process
Seminar	VLSI Layout and Simulation	3rd	18hrs	Backend
Seminar	C-Based Hardware Design	3rd	18hrs	Frontend
Practical	CMOS Gate Array Design and Fabrication	3rd	30hrs	Backend to Process

2.   PRACTICAL COURSE DESCRIPTION

Figure 1 shows Gate Array chips before and after fabrication process. On top of master slice chip, 0.6um-thick aluminum is deposited. The goal of the course is to fabricate and measure Gate Array chips.

Table 2 shows schedule of the course. The course is ten days in total. From first day to fourth day, Design and Simulation is performed. If students successfully finish their design, a teaching assistant fabricates mask on the fifth day with the presence of students. Process and measurement begins at sixth day. Two groups of four students each are accepted for one term of Gate Array practical course. Because of lack of measurement apparatus, two groups cannot measure circuits in the same day. Therefore pipelined schedule is given to the students. The first group processes on the sixth day, then measures in following two days.

A.M. Ionescu et al. (eds.), Microelectronics Education, 139–143.

The second group measures the circuit on the ninth and tenth day. On the sixth day, second group goes into training of measurement by using an example inverter chip. The rest of day is used for literature surveillance or additional simulation to validate the measurement.

Fig. 1: Gate Array master slice (left) and processed chip (right)

Table 2: Gate Array Schedule

Day	Group #1	Group #2
1st - 4th	Gate Array Design	
5th	Visiting In-house Mask Fabrication Facility	
6th	Process	Measurement Preparation
7th	Measurement I	Supplemental Study
8th	Measurement II	Supplemental Study
9th	Supplemental Study	Measurement I
10th	Supplemental Study	Measurement II

2.1   GATE ARRAY DESIGN

The master slice consists of 49 building blocks. A building block contains 3 common-gate PMOSFET and NMOSFET pairs as shown in Fig. 2. Gate length is 3um. Students are allowed to use one metal layer for connection. For crossing, poly-silicon lines can be used as a jumper.

Three CAD tools are supplied through VLSI Design and Education Center's academic program. Cadence Virtuoso CAD tool is used for Layout. Then netlist is extracted using Cadence Dracula. Circuit is simulated with Synopsis Star-HSPICE. Completed gate array design is exported to GDS-II stream format for mask fabrication.

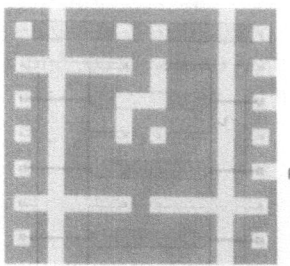

Fig. 2: Gate Array building block

Table 3 shows circuits designed by students in year 2003. A group of two students can either design a large-scale circuit in collaboration, or design the circuit alone. In collaboration students can design a large-scale circuit such as 3bit full-adder with memory. It is very important for them to learn to work together, so that collaboration is highly recommended in this course.

Table 3: Designs submitted in year 2003

2bit full adder	7segment LED decoder
2bit ALU (Adder+Multiplier)	6bit adder
2bit comparator	Boolean function selector
2bit carry look ahead adder	Edge trigger JK-FF
3bit adder with memory	3input AND

2.2   MASK FABRICATION

The in-house mask fabrication apparatus that VDEC has is an electron beam writer (JBX-7000MV). This apparatus can expose very small feature down to 200nm, however, we don't take the maximum advantage of this apparatus; minimum feature size that we limit is 50um. It is mainly because of bad alignment accuracy in photolithography process. Visiting Mask Fabrication Apparatus is optional so that if students didn't finish their layout at the end of fourth day, fifth day is also given for layout and simulation. Whether layout is finished in four days depends on the students; most of the students who take VLSI Layout and Simulation Seminar can finish Gate Array design in four days.

2.3   PROCESS AND MEASUREMENT

Each student executes a standard 1-layer aluminum photolithography and etching process with assistance. After UV exposure on Shipley S1805 photoresist followed by development in NMD-3, aluminum is etched by H_2PO_3 + CH_3COOH +

HNO_3 mixed solution. Figure 3 shows an example of inverter. Minimum width of metal line is 50um and design grid is 25um. The largest discrepancy between design and result is alignment error during exposure. Fabricated gate array chip is measured on a probe station. Signal generator, oscilloscope, and frequency divider (/4, /8, /16) are provided for measurement.

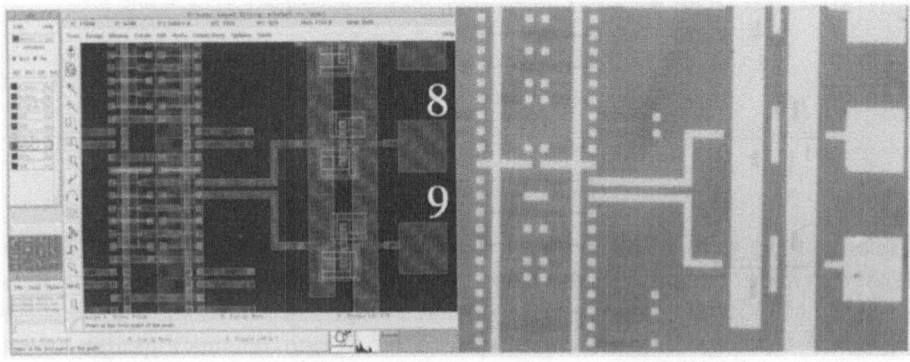

Fig. 3: Design example of inverter (left), fabricated inverter (right)

3.   DISCUSSION ON PORTABILITY

Mask fabrication is the key issue for those who are interested in performing the same kind of practical course. VDEC's commercial-quality EB writer is difficult for universities to obtain. However, since minimum feature size is large, an immersion glass mask can replace the EB mask. Before installation of VDEC's EB, students made immersion mask. Design was printed on an A3-wide paper. Then photograph was taken on a glass plate having black and white immersion layer on top. Necessary apparatus is only a camera for a wide-size film. An example of Immersion mask and EB mask is shown in figure 4. Minimum precision of immersion mask is limited to ten micron because of distortion; that is small enough as compared to allowable alignment error (20um). Since mask fabrication is so simple and standardized at VDEC, EB mask is used instead of immersion mask. It is also unique experience for students to visit the commercial-quality apparatus working for their own design.

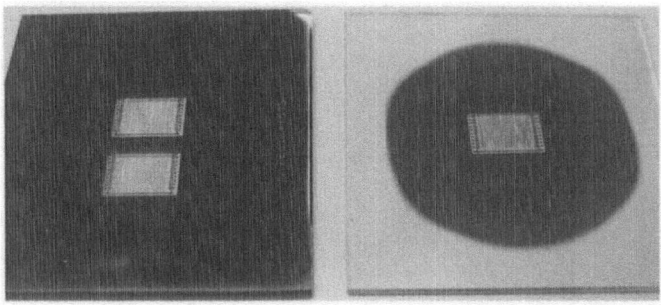

Fig. 4: EB mask (left) and Immersion Mask (right)

4.   SUMMARY

A ten days' practical course titled `Design, Fabrication and Testing of CMOS Gate Array' is given to undergraduate students. Students design one-layer aluminum connections of gate array using Cadence, and make simulations with HSPICE. Then they perform a simple lithography of 0.6um-thick aluminum layer and effectuate measurement with signal generator and oscilloscope. Photomask is fabricated with in-house EB mask fabrication facility.   Each year, maximum 40 students out of 120 total members can take the course.

REFERENCES

[1]   T.Ishihara, S.Komatasu, M.Ikeda, M.Fujita, and K.Asada, "Comparative Study on Verilog-based and C-based Hardware Design Education", In the Proceedings of the 2003 MSE conference, Anaheim, USA.

NANOTECHNOLOGIES AND NANOELECTRONICS

Lectures and practical works at the nanoscale in Grenoble
MONTES L., CLERC R., IONICA I., BALESTRA F., TSAMADOS D., PERNOT E., CORNU S., SCHAEFFER C.
Institut National Polytechnique de Grenoble (INPG), 46 Av. Félix Viallet, 38031 Grenoble Cedex 1, France
MARCHI F., CHEVRIER J.
Université Joseph Fourier (UJF), Campus Universitaire, 38042 Saint Martin d'Hères, France

1.   INTRODUCTION

Partly thanks to the evolution of microelectronics, nanoscience is now a fast evolving field of modern science. As many universities in Europe, the two universities of Grenoble, the Université Jospeh Fourier (UJF) and the Institut National Polytechnique, have engaged many actions to offer master degree syllabi including or requiring the knowledge and use of micro- and nano- technologies (MNTs), in order to respond to the growing demand of industry, research centers and laboratories.

In this paper, we present a new course devoted to MNTs and related labworks.

2.   LECTURES

Optional lectures are proposed to last year engineering students (master degree) of two engineering schools in Grenoble: ENSERG (electronics) and ENSPG (physics), concerning 23 students in 2002-03 and 50 in 2003-04. This course presents the actual research and foreseen elaboration techniques, architectures, devices and characterization methods in nanoelectronics. The objective is to give the students an overview of MNTs. The first part concerns advanced microelectronics: ultimate MOS architectures (multi-gate, SON, vertical, …) SOI technologies, and related problems. Then nanotechnologies are detailed through the top-down and bottom-up approaches, and self-organization. Metrology and characterization techniques are mainly evidenced with near-field instrumentation and analysis. The specific nanoelectronic phenomena are introduced: tunnel transport, quantum confinement, Coulomb blockade. These effects are illustrated through monoelectronic devices: single electron transistors (SET) and memories (SEM). A particular emphasis is made on related problems such as technological fluctuations, offset charges. Fault tolerant and specific architectures are enounced.

A.M. Ionescu et al. (eds.), Microelectronics Education, 145–149.

Tunnel junction devices (interband tunneling diode, resonant tunneling devices), wave interference devices, spintronics and few optronic devices are also evocated. Finally, molecular and polymer electronics is proposed as possible alternative solutions.

3.   FROM MICRO TO NANO ELECTRONICS LABWORKS

On the practical side, we have set up a new silicon process flow using 0.6µm technology, compared to 2µm technology actually in use in CIME[1]. The gate oxide is reduced from 50 to 20nm, a 250nm deep-UV photolithographic system (Figure 1) is used, the annealing changed to rapid thermal annealing (Figure 2).

The layout (Figure 3) consists of elementary devices such as diodes, MOS capacitors and transistors, and many test structures as Transmission Line Method, Hall Effect and Van Der Pauw. Specific transistors in 0.25µm industrial technology are also used for comparing the electrical characteristics[2] and introducing specific effects (SCE, DIBL, …).

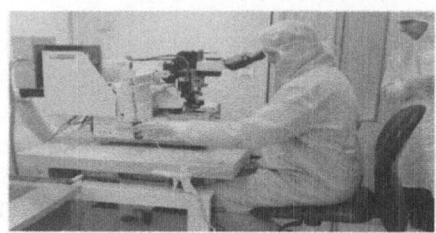

Fig. 1: DeepUV lithography

Fig. 2: Rapid Thermal Annealing

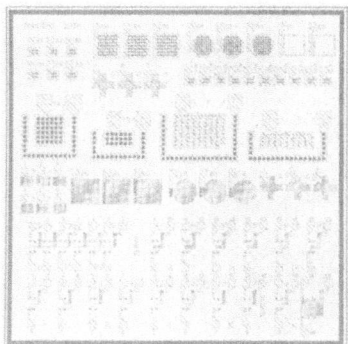 

Fig. 3: Example of 2 chips layouts (gresne and kal)

4. THE NANO PLATEFORM IN CIME

A new program to setup a technological plateform dedicated to teaching nano-technologies and nanoelectronics is now started in CIME. It is supported by the well established local scientific community in Physics and Microelectronics, and the numerous researchers and professors working in the field of nanotechnolo-gies. This platform is financially supported by the scientific universities (INPG, UJF) and engineering schools of Grenoble, and the Region Rhones-Alpes, as it is a complement of the existing trainings in microelectronics and microsystems[3]. It is composed of several microscopes using different approaches: atomic force (AFM in Figure 4), tunneling current (STM capable of obtaining atomic resolution as in Figure 5), optical interference, electron beam microscope, mechanical pro-filometers. This platform has no equivalent in France and should be used by 200 students/year, at the master or doctorate degree in Physics or Electronics. These instruments will be the basic and common tools to approach two topics: instru-mentation and physics on one side and nanoelectronics on the other side. The different labworks will be focused on near-field instrumentation, nanomechanics, nanotribology, nanofabrication (AFM oxidation, nanoimprint, nanomanipulation). For the nanoelectronics topic, the general idea is to set up a series of five lab-works to fabricate, characterize and simulate (Figure 6) nanodevices.

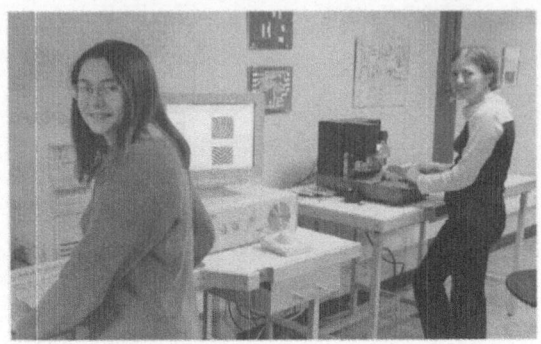

Fig. 4: Atomic Force Microscope (AFM)

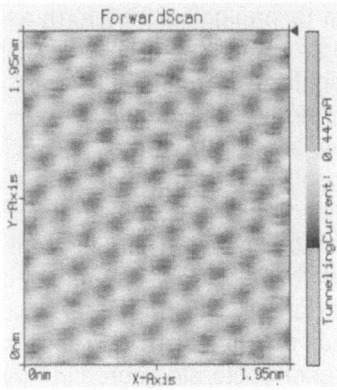

Fig. 5: Atomic resolution obtained on graphite surface using a STM

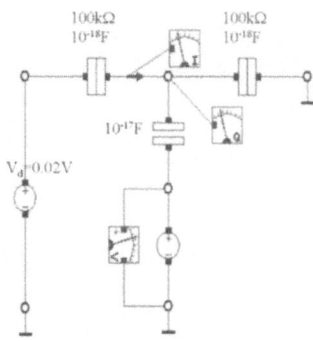

Fig. 6: SET simulation using Simon software

6. CONCLUSIONS

To complete the Minatec project in Grenoble and the expanding syllabi in local universities that include nanoscience education, a new lecture and a technological plateform in nanotechnologies have been created. These actions should help promoting the micro- and nano- technologies and anticipate the demand of industry and laboratories. They will be used in new trainings, as the 'Nanoworld Project' oriented to high school students[4], the European School on Nanoscience-Nanotechnology[5], and an international master on MNTs.

We are grateful to the engineers and technicians in CIME and the many researchers who helped in setting up the technological plateform.

References

[1]   Centre Interuniveristaire de MicroElectronique de Grenoble, UJF-INPG clean room facilities

[2]   L. Montès, L. Palun, N. Mathieu, P. Morfouli, A. Pouydebasque, N. Guillemot, C. Schaeffer, J.M. Terrot, C. Uzel, B. Gonzales, I. Pheng, F. Parrain, EWME2002 proceedings, p.9-12

[3]   S. Haendler, L. Montès, N. Mathieu, P. Morfouli, N. Guillemot, F. Parrain, B. Charlot, EWME2002 proceedings, p.121-124

[4]   See the abstract on 'Nanoworld Project', in the same EWME04 volume.

[5]   http://www.esonn.inpg.fr

POSTER PRESENTATIONS

COMPUTER AIDED ANALYSIS OF THE PARASITIC PROPERTIES OF A BIPOLAR TRANSISTOR CELL

DONOVAL D., CHVALA A., VRBICKY A.
e-mail: daniel.donoval@stuba.sk
Department of Microelectronics, Slovak University of Technology in Bratislava,
Ilkovicova 3, 812 19 Bratislava, Slovakia

Abstract

The computer aided analysis comprising TCAD process and device simulation as well as SPICE modeling and simulation of electrical properties of a bipolar transistor cell for analysis of experimental results is presented. A unique insight into the internal bipolar transistor structure operation allows for better understanding of electro-physical behavior of semiconductor structure and subsequent extraction of SPICE equivalent circuit model including parasitic devices. Very good agreement of simulated electrical characteristics with experimental ones confirms the validity of the derived model. The implementation of complex simulation tools for structure, device and circuit simulation into microelectronics curricula, particularly for students projects allows for better understanding of IC and systems.

1.   Introduction

The growing complexity and increasing on-chip circuit and system integration enhanced by continuing miniaturization of individual semiconductor devices, which are approaching their physical limits, generate a strong pressure on microelectronics curricula. Simultaneous use of process and device simulation allows the deep understanding of physical behavior of operation of basic semiconductor devices and building blocks and provides the designers and system engineers with necessary background to understand the circuit operation, extract parasitic effects and exploit maximum performance from a given technology [1,2]. The massive implementation of advanced design software tools for circuit and system analysis, methods of design, modeling and simulation, characterization, verification and testing of new technologies may also help to attract more students to microelectronics providing the easier way of understanding of the non-easy complex phenomena which is considered as one reason of the inadequate low interest of young generation in a difficult microelectronics study [3,4].

An example of the implementation of TCAD process and device simulation of the n-p-n bipolar transistor with buried collector and reverse biased p-n junction isolation in the common emitter configuration is presented. It allows the interpretation of the electrical characteristics by the knowledge of internal electro-physical properties of the structure. Based on the better understanding of the internal behavior of transistor structure the equivalent circuit model was derived and its approximative parameters were extracted and used for SPICE simulation. The

153

A.M. Ionescu et al. (eds.), Microelectronics Education, 153–158.

obtained good qualitative agreement of the simulated results of circuit simulation by SPICE, numerical process and device simulation by TCAD and experimental results confirm the validity of the derived model. The wide implementation of complex SPICE and TCAD tools for laboratories and practical student training in microelectronics curricula allows students to understand the internal electrophysical behavior and extract the parasitic effects which importance is strongly increasing with decreasing the structure dimensions. The students are then very well prepared for IC and system design and analysis not only on a circuit level but also on the device and structure level, they can identify and analyze the critical regions and parasitics responsible for device false operation and failure.

2.   Device structure

The typical structure of the NPN bipolar transistor with buried collector and reverse biased PN junction isolation which corresponds to the real bipolar IC technology was designed and simulated by 2-D process simulation using ISE TCAD process simulator DIOS [5] (Fig. 1). The N-type epitaxial layer was grown on p-type substrate into which the N^+-type buried collector and P^+-type buried region for p-type isolation ring were implanted prior to epitaxial growth. The ion implantation steps in predefined regions consisting P-type isolation ring, P-type base and N^+-type emitter and collector planar contact followed by high temperature activation of impurities create the bipolar transistor structure as basic building block of bipolar IC technology. Planar ohmic contacts to emitter, base, collector and substrate for applying external voltages were defined.

3.   Experimental results

The very good qualitative agreement of the simulated and measured static I-V characteristics of the model structure of bipolar NPN transistor in the common emitter configuration, namely base, collector, and substrate currents I_b, I_c, and I_s and extracted value of the common emitter current gain β are shown on Fig. 2.

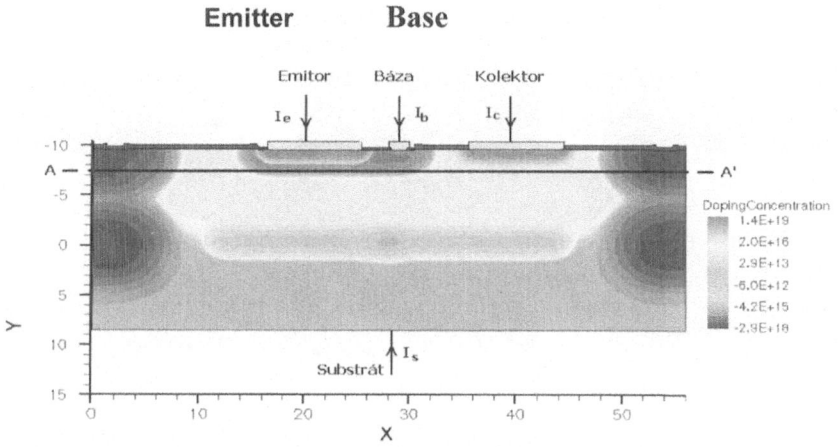

Fig. 1: Schematic of the structure of bipolar transistor with buried collector and reverse biased PN junction isolation

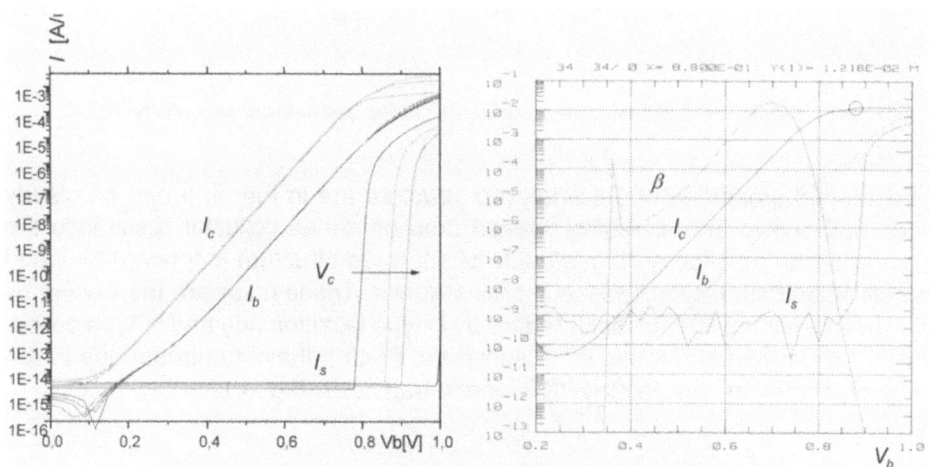

Fig. 2: Simulated and measured Gummel plot of bipolar transistor

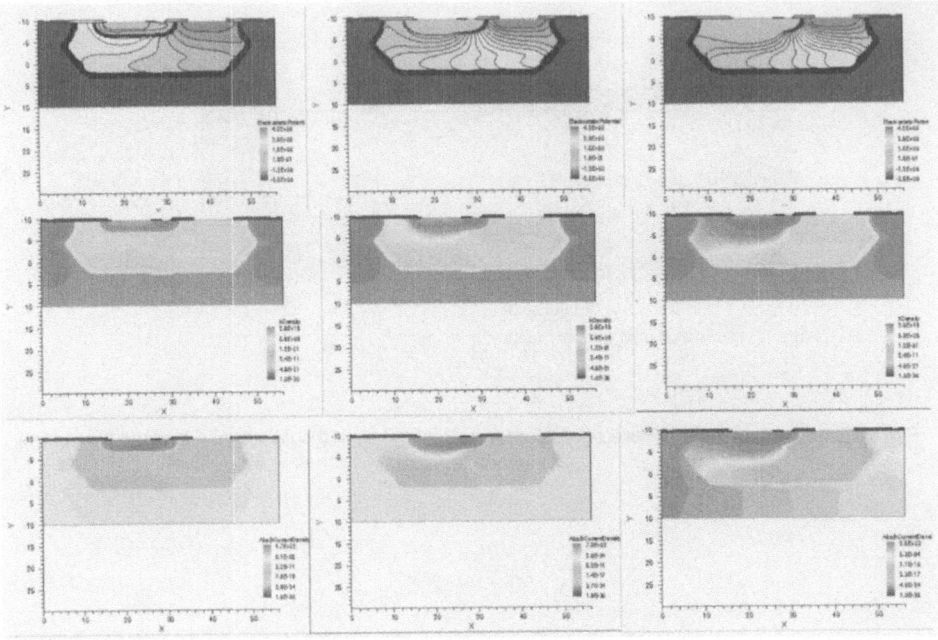

Fig. 3: The distributions of potential, free holes, and hole current density

The internal properties of the analyzed structure are in Fig. 3. It can be clearly seen that due to an increasing voltage drop on series collector resistance the base-collector junction on the left side of analyzed structure is forward bias and injects holes from P-type base to N-type collector. These holes are then swept by the reverse biased PN junction created by P-type isolation ring and N-type collector and large hole current flows to substrate. Such behavior corresponds to the creation of the parasitic lateral PNP bipolar transistor (Fig. 4.).

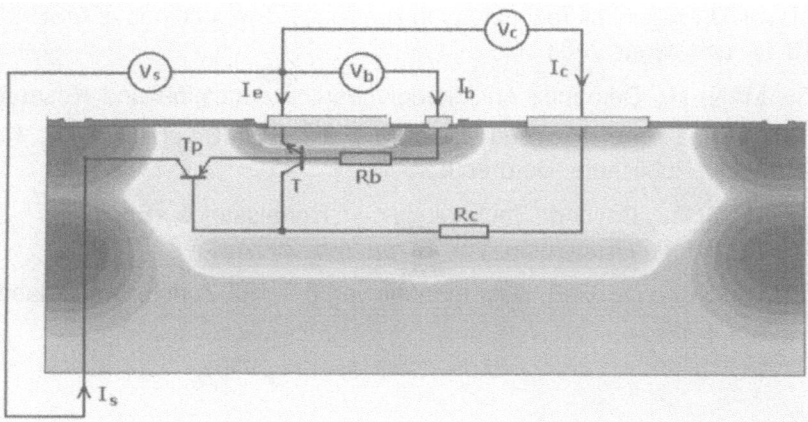

Fig. 4: Equivalent circuit model with the parasitic lateral PNP bipolar transistor

Conclusion

The 2-D numerical process and device modeling and simulation is presented as the very effective tool for the better understanding of internal behavior of complex semiconductor structures and so for the interpretation of experimental electrical characteristics of equivalent semiconductor structures. The visualization of internal electro-physical properties of analyzed structure which cannot be directly measured as potential, carrier densities, electron and hole currents, impact ionization and others and their correlation with output electrical characteristics is extremely useful for physical interpretation of obtained data. The user with some expertise can analyze the critical points and regions and extract the parasitic effects. The simultaneous use of SPICE modeling and simulation together with TCAD process and device simulations provide creative environment for young experts. They will be very well prepared to solving the complex problems of advanced semiconductor structures not only on the circuit but also structure and device levels.

Acknowledgement.

This work was supported by grant APVT-20-013902 and KEGA 53 of the Slovak Ministry of Education.

References

[1] BRADLEY, B. S., Principles vs. Practices in Undergraduate Microelectronics Systems Education, In: Proceedings of MSE'01, pp. 22-23, Las Vegas 2001

[2]   DONOVAL, D., An Industrial Impact on the Microelectronic Education at Slovak University of Technology in Bratislava, Proceedings of MSE'01, pp. 16-17, Las Vegas 2001

[3]   DE MAN, H., Demands on Microelectronics education and Research in Post – PC Area, Proceedings of the 3rd EWME, pp. 9 - 14, Kluwer Academic Publishers, Dordrecht 2000

[4]   RAINEY,V.P., Beyond Technology – Renaissance Engineers, IEEE Transactions on Education, Vol. 45, pp. 4-5, 2002

[5]   ISE DIOS and DESSIS, User manual, ver. 8.0, ISE Zurich, Switzerland

EMERGING NANOELECTRONICS: LIFE WITH AND AFTER CMOS

BANERJEE K. (1), IONESCU A.M. (2)
(1) University of California, Santa Barbara, (2) LEG, Swiss Federal Insitute of Technology, Lausanne, Switzerland

1.   Why do we need this Book?

At present, *nanoscale effects* in CMOS and their design implications, as well as emerging nanoelectronic devices and circuit architectures, are not adequately covered through university (undergraduate and graduate) curricula.   However, with CMOS dimensions scaling steadily below 100 nm, nanometer scale effects (including quantum effects) are becoming so significant that even the random placement of dopant atoms in the channel region of a MOSFET can cause undesirable device characteristics. For emerging nanotechnologies, these effects are expected to increase even further.   Hence, there is an immediate need in academic world to build a strong foundation geared towards addressing *nanoscale technology* issues to prepare today's students for the coming generations of integrated circuit design. Unfortunately, there are no text books that comprehensively cover this subject. The aim of this book is to bring together some of the most significant works reported on *nanoscale CMOS* and *emerging nanoelectronics*, to provide the reader with a comprehensive overview of what the challenges are for nanoscale CMOS and what technology options exist beyond CMOS. In that regard, this book is intended to provide an excellent "one-stop" information package on nano-CMOS and emerging nanotechnologies. It can not only serve as the main information source for a graduate course on nanoelectronics, but can also be used as an excellent reference for advanced graduate level classes in device physics or VLSI circuit design. We believe that its content would be of high interest not only for students but also for engineers, scientists and researchers who are naturally migrating towards the 'nano' world.

2.   How did this Book Emerge?

The concept of preparing this book emerged from our efforts to comprehend some basic questions about nanotechnology: what does it mean? Where is it heading? Can it be a good candidate to replace or perhaps combine with CMOS in the future? Will it revolutionize our lives? Based on some interesting discussions during the summer of 2001, a time we worked together in EPFL-Lausanne, we decided to prepare such a manuscript, primarily to clarify our own understanding of the 'nano' world and to identify new research directions. Our ongoing collaborations in nanoelectronics over the past three years have also helped shape the contents of this book. Furthermore, parts of the book resulted from our efforts

159

A.M. Ionescu et al. (eds.), Microelectronics Education, 159–160.
© 2004 *Kluwer Academic Publishers.*

in preparing class material for a graduate level course: ECE594K, *'Special Topics in Nanometer Scale VLSI'* offered at UCSB and *'Devices and Circuits of the Future: Towards Quantum Electronics'* at EPFL.

3.   What does this Book Contain?

The book is a collection of papers that aims to introduce the reader in the emerging nanoelectronics world, structured as follows:

- 3 introductory papers (including the famous talk given by Richard Feynman in 1959: "There is Plenty of Room at the Bottom")
- 61 papers on Nano-CMOS
- 18 papers on few-electron devices including Single Electron Transistor and Single Electron Memory
- 20 papers on emerging and ultimate memory architectures
- 34 papers on nanoscale circuits
- 19 papers on emerging nanotechnologies

MODEL ELECTRONICS PROGRAM AT UNIVERSITY OF MASSACHUSETTS LOWELL

PRASAD K.

Ph.D.; P.E. Professor, Electrical and Computer Engineering Director, Microelectronics / VLSI Technology University of Massachusetts Lowell, USA
Kanti_Prasad@uml.edu

Abstract

In 1986, Microelectronics / VLSI Technology program was established at UMASS Lowell, as a result of proposal submitted by the author to the Massachusetts Microelectronics Center (MMC). One million dollar funding in kind and cash was matched by Chancellor Hogan in converting an existing space of 2400 sq feet in to a modern clean room facility. Two Laboratories namely: VLSI Design Lab and Distributed Semiconductor Industrial Processing Laboratory (DSIPL) were established deploying the state-of-art design and fabrication facilities.

In order to carry out consummate instructions in Microelectronics / VLSI Technology field, author envisioned development of state-of-art curricula, supported by the state of art Laboratories, in order to design, test and fabricate microelectronic circuits. The courses such as 16.502 VLSI Design, 16.602 VHDL Based Design, 16.504 VLSI Fabrication and 16.574 MMIC Design and Fabrication were developed and taught by the author all along.

In order to comprehend the microelectronics field in its entirety, the author envisioned a set of four labs from the date of establishment of this Microelectronics Center at UMASS Lowell, namely: (1) VLSI Design Lab. (2) Mask Fabrication Lab. (3) DSIPL, and (4) Dicing and Packaging Lab. The (1) and (3) are achieved in its entirety, (4) is partially achieved, but (2), still needs to be achieved. In the meanwhile, we are getting MASK set from National Institute of Science and Technology (NIST). It is because of this reason, we are capable of imparting state-of-art instructions in the field of Microelectronics VLSI Design Lab incorporates the latest Cadence software bundles along with Solaris-8 implemented on Sun Sparc-10 stations.

The author however emphasized the Microelectronics Model [1] comprising of (1) Fundamentals, (2) Materials, (3) Devices, (4) Circuits, and (5) System and is tirelessly working in preparing students for the Hi-Tech Industry, based on this model. Although the articles were primarily dedicated to educational endeavors, but these have resulted in sizable research funding also, primarily from regional Hi-tech industry.

Reference

[1]   Prasad, K., "Development if Microelectronics Engineering Education Model and its Deployment in Preparing the Technical Manpower for the 21st

A.M. Ionescu et al. (eds.), Microelectronics Education, 161–162.

Century", published in Canadian Procedures of Engineering Education, 1998.

CMOS DEVICES LIFETIME PREDICTION METHOD

MONGELLAZ B.
*IXL Laboratory, Bordeaux 1 University, 351 cours de la libération, 33405
TALENCE, France*
email : mongellaz@ixl.fr

1.   Introduction

Hot carrier induced degradation of MOSFET performance versus time is an important reliability concern in modern microcircuits. The development of reliability tools which aim to predict the degradation in circuit performance after a specified operating period is also a considerable interest within the semiconductor industry [1]. The concept of Design For Reliability (DFR) is a new challenge. Design For Reliability strategy is based on reliability simulation in the design flow that aims to predict the impact of physical phenomena from MOSFET to IC electrical characteristics over the operating time. Consequently, reliability tools need to take into account reliability models of transistor. The challenge for reliability simulation is to link the physical degradation mechanisms on the electrical characteristics of an electronic system from the transistor level to the circuit level or system level [2]. The major requirement in term of reliability simulation is to calibrate transistor device reliability model based on experimental case study [3]. A synthesis of such transistor device reliability models is proposed concerning CMOS process technology versus a wear-out failure : hot-carrier injection. A discussion is introduced about how using these device reliability models to analyse circuit level reliability. This paper is an approach dedicated to masters student people to get knowledge to better apprehend semi-conductor device reliability by electrical simulation at the transistor level. This work could be included in microelectronics education course focusing on semi-conductor device reliability in complementary of integrated circuits design.

2.   A wear-out failure mechanism

The two major hot-carrier injection mechanisms for a NMOS are activated by the lateral electric field. It accelerates carriers to sufficiently high energy so that they can cause impact ionisation through collisions with the silicon lattice near the drain end. The initial electron-hole pair creation can cause further impact ionisation leading to an avalanche effect. The resulting mechanism is known as drain avalanche hot-carrier injection. The generated carriers flow out of the substrate, which gives rise to a hole substrate current in an NMOS and an electron substrate current in a PMOS as shown by Figure 1. If the channel electron has sufficient energy to overcome the S_iO_2/S_i barrier, it can be redirected and injected into the

A.M. Ionescu et al. (eds.), Microelectronics Education, 163–168.
© 2004 *Kluwer Academic Publishers.*

gate oxide. This mechanism is known as channel hot-electron injection. The hot-carriers near the drain can also overcome the S_iO_2/S_i barrier resulting in a predominantly electron gate current for both NMOS and PMOS. These hot-carrier injection mechanisms have been found to cause device damage. Depending on the operating condition, one or more of these hot-carrier injection mechanisms can take place in a single transistor. The injected hot holes and electrons can cause oxide and interface damage by creating traps in the oxide, being trapped in the oxide, creating interface states at the interface, and or being trapped by interface states. At the transistor level, manifestation of oxide and/or interface damage can often be observed as a shift in threshold voltage and a decrease in carrier mobility. For NMOS, the threshold voltage typically increases due to predominant electron trapping, and this increase coupled with mobility decrease results in a smaller drain current. For PMOS, however, the electron-trapping results in threshold voltage decrease in absolute value, causing the drain current to increase despite decrease in mobility. All this injection mechanisms activation highly depend on voltage stresses applied to CMOS devices.

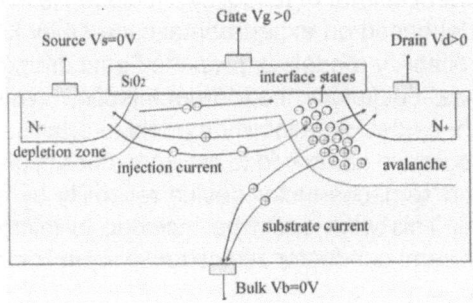

Fig. 1: Hot-carrier injection in a NMOS semiconductor

3. CMOS devices lifetime prediction models

For measuring the hot-carrier induced degradation, accelerated tests are performed using a single MOSFET devices. The method recommends for accelerated experiments to increase the drain-source voltage only in a range from 1.1Vdd (maximum process power supply) to 90% of snapback voltage. Or in other case, the gate-source voltage can be calibrated to bias CMOS device in a worst case under peak substrate current. This mechanism occurs most easily when V_{GS} is approximately $0.5V_{DS}$. All this experiments are performed along several stress cycle time. Each device is characterized at the end of each stress interval. The transfer characteristics $I_{DS}=f(V_{GS})$ and the output characteristics $I_{DS}=f(V_{DS})$ are used to extract CMOS stressed electrical parameters set. The evaluation of

CMOS device lifetime is based on the parameter shifts measured during DC stress tests. A CMOS device under test is considered to fail when a chosen device parameter changes by more than the specified failure criterion at a specified end of life corresponding to ten years operating life. These failure criteria for the monitored parameters depend on the specific requirements to the technology. Degradation criteria commonly used are 10% change in saturation drain current I_{DSsat}, linear drain current I_{DSlin} or 100mV threshold voltage V_{TH} shifting. These information are obtained by experiments and can be modelled in accordance with three commonly used models. These models take in account bias dependence and are useful to estimate target degradation t_d. Each model have fitting parameter dependent on the technology under consideration. The first model, called Takeda model, depends on drain-source voltage that is a preponderant bias condition to accelerate hot-carrier injection [4] :

$$t_d = t_0 \exp(\frac{B}{V_{ds}}) \text{ (sec)}(1)$$

The second model, called Hu model, is from substrate-drain current ratio method [5]. This model takes in account W the channel width, the drain current I_{DS} and the substrate current I_B that is correlated to gate-source bias voltage chosen to have MOSFET device under peak substrate current worst case :

$$t_d = \frac{HW}{I_{DS}} \left(\frac{I_B}{I_{DS}}\right)^{-n} \text{ (sec)}(2)$$

These models proposed are also available to predict PMOS lifetime. In this case, substrate current is changed by gate current using Hu reliability model. These models above are defined as standard model that are commonly used to predict CMOS device lifetime under experiments. These three models are available for extrapolation of experimental results to nominal bias conditions. The Figure 2 shows an example using drain-source voltage acceleration method. MOSFET transistors are stressed by high V_{DS} voltages and nominal V_{GS} voltages. The failure criterion is 100mV threshold voltage V_{TH} shifting and lead to the corresponding lifetime. Then, the CMOS lifetime due to operating bias conditions are extrapolated. Moreover, the drain-source voltage induces a ten years lifetime is defined.

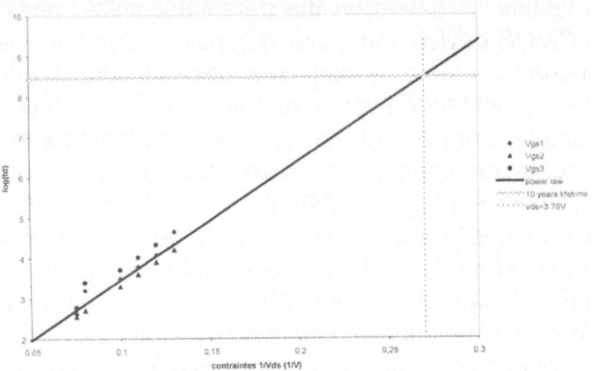

Fig. 2: NMOS lifetime prediction using Takeda model

4.   Lifetime prediction by simulation

An application to predict CMOS device lifetime at the design phase is proposed. The method consists to include the reliability models explained above in an electrical simulation environment. The ELDO electrical simulator is chosen to perform lifetime prediction by simulation. This electrical simulator propose a great choice of CMOS device model. The CMOS device model used during the design phase is the BSIM model. This electrical device model is well-defined since it takes into account substrate current model and gate current model. That means all these electrical quantities that define in reliability model are available. BSIM device model is consequently a convenient MOSFET device model to use in order to demonstrate the feasibility to include reliability analysis during design phase. The goal is to demonstrate that designer can get information about device lifetime under design phase by applying a specific reliability model. The electrical simulation is performed using ELDO simulator. Since a specific simulation analysis as operating point simulation can give information about bias voltage or current terminal devices. The prediction of CMOS lifetime is automatically performed. These simulation data are vital to the designers because they can locate CMOS transistors that lead to shorter lifetime. Shorter lifetime means that the electrical performance of MOSFET devices can degrade faster over time. Moreover, these respective MOSFET can be weak spot of integrated circuits. Thereby, the integrated circuit performance can be reduced by these stressed devices over operating time. Consequently, these simulation results are exploited to improve the topology of integrated circuits.

The application is based on CMOS process technology dedicated to analogue integrated circuits. The demonstrator circuit is a basic analogue block: an operational transconductance amplifier (OTA). The OTA is designed using a 0.8μm CMOS technology. The drain-source acceleration method is applied to define a respective Takeda reliability model. The simulation results obtained locate the MOSFET transistor that has sensitive to drain-source stress voltage. The evaluation of each MOSFET transistor lifetime shows that the architecture of the OTA need to be improved by adding specific MOSFET transistor to reduce drain-source bias voltage. For example, the output stage that defines the output resistor of the operational amplifier is a weak spot. A solution to reduce drain-source voltage of each MOSFET transistor is to modify the topology of the output stage. A cascode output stage is preferred. Moreover, this solution increases the OTA output resistor.

This application will be develop in details including specific data to demonstrate the feasibility of CMOS lifetime prediction during design phase. Thereby, this application can be extent to other analogue integrated circuits. The main goal is to show the necessity to include this concept in design tools in order to give opportunity to designers to evaluate the reliability of integrated circuits. Of course, this application is first proposed to masters student people to show how apprehending the CMOS device reliability by experiments and by electrical simulation.

Conclusion

Reliability simulation is in mature phase. Reliability models are defined based on standard experiments method. These reliability models are now included in design tools as proposed by the paper. The application focuses on CMOS lifetime prediction versus hot-carrier injection. Performance degradation modelling induced by hot-carrier injection is very interesting for semiconductor factory mainly to future process technology. Reliability analysis by electrical simulation of integrated circuits is great concern for microelectronic industry. Consequently, a fresh activity needs to be developed and it is really important to prepare student people to reliability of semiconductor devices. This challenge could be manage by including reliability tools in microelectronics education as proposed in this paper. These tools have to be introduced to student people to develop new professional skills. Included in integrated circuits design course, this purpose leads to a complementary view of design phase, called design for reliability.

References

[1]   Matthewson A., "Modelling and simulation of reliability for Design," in *Microelectronics Reliability*, 1999, pp. 95-117.

[2]   B. Mongellaz, F. Marc, N. Lewis, Y. Danto, "Contribution to ageing simulation of complex analogue circuit using VHDL-AMS behavioural modelling language," in *Microelectronics Reliability*, 2002, pp. 1353-1358.

[3]   JEDEC standard JESD-28, "A procedure for measuring N-channel MOSFET hot carrier induced degradation at maximum substrate current under DC stress,", june 1995.

[4]   E. Takeda, N. Suzuki, "An Empirical Model for Device Degradation Due to
      Hot-Carrier Injection," in *IEEE Electron Device Letters*, vol. edl 4, No. 4,
      april 1983.

[5]   C. Hu, "IC Reliability Simulation," in *IEEE J. Solid-State Circuits*, vol. 27,
      No. 3, march 1992.

ANALOG INTEGRATED CIRCUIT OPTIMIZATION USING SPICE

SERRA-GRAELLS F.[1], URANGA A.[2], BARNIOL N.[2]
[1]*Centre Nacional de Microelectrònica, CSIC*
[2]*Electronic Engineering Dept., Universitat Autònoma de Barcelona*
Campus UAB, 08193 Bellaterra, Spain

1.   INTRODUCTION

The increasing market demand on mixed systems-on-chip (SoCs) requests electronic engineers with good knowledge on both, digital and analog integrated circuits. Furthermore, the design of the analog blocks in these SoCs is usually more time consuming and harder to optimize than their digital counterparts. This paper proposes a training strategy based on Spice simulation to teach analog integrated circuit optimization. The choice of the Spice-language standard ensures the applicability of the acquired knowledge in most of the analog design fields for SoCs (e.g. signal processing, RF communications, power control). In this sense, the Spice Opus CAD tool [1] have been selected here, which is a free Berkeley Spice3 compliant simulator with optimization and XSpice [2] mixed-mode capabilities, so it can be also reused for other related syllabus like full-custom design [3] or mixed modelling [4]. In order to illustrate the proposed methodology, a circuit case study is fully developed in the following sections as a real practice example for students.

2.   TEST CIRCUITS

The practice example presented here is built around the classic dual-stage Miller operational amplifier of Figure 1(left). In this case, the main specifications are: differential voltage gain at DC (gdc), gain-bandwidth product (gbw), phase margin (pm), positive/negative slew-rates (srpos/srneg), Si circuit area (area) and power consumption (pd). The first step for the student is to model this cell under test in terms of a Spice subcircuit, like in Figure 1(right). As a CMOS circuit, the basic design variables of this cell are the device dimensions (i.e. w and l) and bias, which in fact are related to the specifications area and pd, respectively.

A.M. Ionescu et al. (eds.), Microelectronics Education, 169–174.

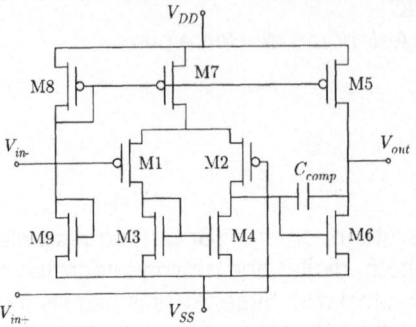

```
.subckt opamp vinn vinp vout vdd vss
m1 vload vinn vcomm vdd modp w=18u l=6u
m2 vinter vinp vcomm vdd modp w=18u l=6u
m3 vload vload vss vss modn w=12u l=12u
m4 vinter vload vss vss modn w=12u l=12u
m5 vout vbias vdd vdd modp w=96u l=6u
m6 vout vinter vss vss modn w=48u l=6u
m7 vcomm vbias vdd vdd modp w=24u l=6u
m8 vbias vbias vdd vdd modp w=12u l=6u
m9 vbias vbias vss vss modn w=3u l=24u
ccomp vout vinter cpoly w=100u l=100u
.ends
```

Fig. 1: Operational amplifier under test (left) and equivalent Spice sub-circuit (right)

In order to cover all the required specifications, the student has to manage with several test circuits around the cell. A good solution in this case is the double test set-up shown in Figure 2, where `gdc`, `gbw` and `pm` specifications can be computed from the quasi open-loop topology, while `srpos/srneg` are measured using the follower scheme. Finally, the static `area` and `pd` values may be derived from an operating point in either configuration.

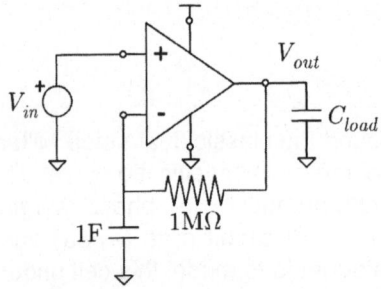

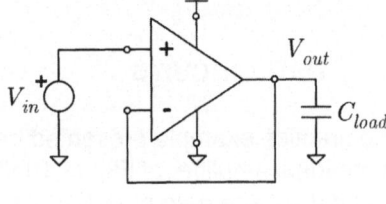

```
qopenloop.sp3
.include cnm25typ.mod2
.include opamp.sp3
cload vout 0 10p
vin vin 0 dc=2.5v ac=1
xopamp vfb vin vout vdd vss opamp
vss vss 0 dc=0v
vdd vdd 0 dc=5v
cdc vfb 0 1
rdc vfb vout 1000k
.end
```

```
follower.sp3
.include cnm25typ.mod2
.include opamp.sp3
cload vout 0 10p
vin vin 0 pulse(2 3 1u 1n 1n 2u)
xopamp vout vin vout vdd vss opamp
vss vss 0 dc=0v
vdd vdd 0 dc=5v
.end
```

Fig. 2: Quasi open-loop (left) and follower (right) test circuits

3.   AUTOMATIC CHARACTERIZATION

The next step towards the cell optimization is the automatic extraction of its performance. The challenge for the student in this stage is both, to perform several analysis in different test circuits and to program the automatic measurements to be applied to the simulation results. A simplified example is listed as follows:

```
.control
source qopenloop.sp3
source follower.sp3
setcirc qopenloop.sp3
op
let pd=-i(vdd)*v(vdd)*1e3
let area=(@m1:xopamp[w]+6u)*(@m1:xopamp[l]+11u)+(@m2:xopamp[w]+6u)*(@m2...
ac dec 50 10 10e6
let gmag=20*log10(mag(v(vout)))
let gph=phase(v(vout))
let gdc=gmag[0]
let gbw=abs(frequency[sum(gmag ge 0)])/1e6
let pm=180+gph[sum(gmag ge 0)]

setcirc follower.sp3
tran 1n 5u
let vrise=v(vout)*(time lt 3u)+3*(time ge 3u)
let trisebot=min(time*(vrise ge 2.1)+(vrise lt 2.1))
let trisetop=min(time*(vrise ge 2.9)+(vrise lt 2.9))
let srpos=0.8/(trisetop-trisebot)*1e-6
let vfall=v(vout)*(time ge 3u)+3*(time lt 3u)
let tfalltop=min(time*(vfall lt 2.9)+(vfall ge 2.9))
let tfallbot=min(time*(vfall lt 2.1)+(vfall ge 2.1))
let srneg=0.8/(tfallbot-tfalltop)*1e-6
.endc
```

After applying such a script to the initial cell design of Figure 1(right), the extracted datasheet compared to the simulation results are both displayed in Figure 3.

Fig. 3: Simulation results (center and right), and datasheet (left) after automatic extraction.

4.   OPTIMIZATION PROCESS

Once the cell under test can be automatically evaluated, the student begins the optimization definition itself, which involves the declaration of the design variables, the required analysis, the implicit constraints, and the cost function. In this sense, the student should minimize the number of design variables by, for example, exploiting circuit symmetries. Also, proper ranges and initial values should be derived from the analytical design equations. In the case of the cell example of Figure 1:

```
optimize parameter 0 element m1:xopamp parameter w low 12u high 200u initial 12u
optimize parameter 1 element m5:xopamp parameter w low 12u high 200u initial 96u
...
optimize analysis 0 setcirc qopenloop.sp3
optimize analysis 1 op
...
optimize implicit 0 op1.pd lt 1.5
optimize implicit 1 op1.area lt 0.025
...
```

Finally, but most important, the success of any optimization process is not only given by the algorithm itself but also by the cost function, which classifies the solution obtained in each iteration. At this point, students can experiment with a wide variety of expressions and take experience from their results, as shown in Figure 4.

```
optimize cost 8*abs((20-tran1.sr
neg)/tran1.srneg)+8*abs((20-tran
1.srpos)/tran1.srpos)+10*abs(tra
n1.srneg-tran1.srpos)+300*abs((8
0-ac1.gdc)/ac1.gdc)+11*abs((10e6
-ac1.gbw)/ac1.gbw)+65*abs((op1.a
rea-0.025)/op1.area)+300*abs((ac
1.pm-50)/ac1.pm)+250*(abs(op1.pd
-1.5)/op1.pd)
optimize cost 1/abs(ac1.gdc)


optimize cost 1/abs(ac1.gdc)
```

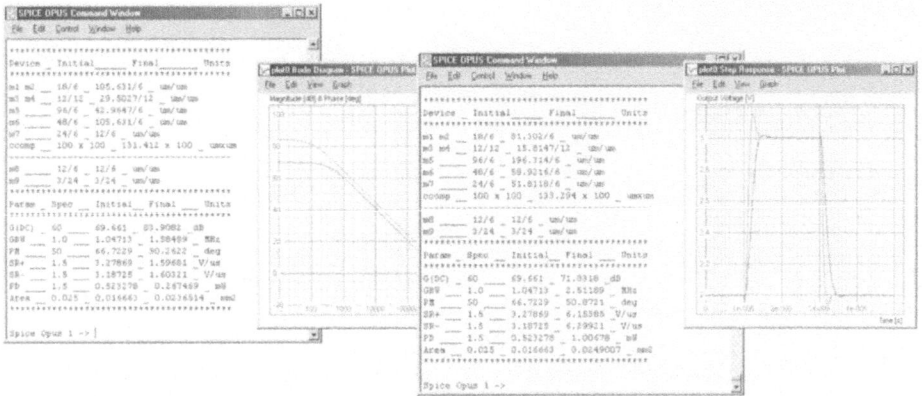

Fig. 4: Cost function example and results for both, global (upper) and specific (lower) optimization

Conclusions

A training methodology for analog integrated circuit optimization has been proposed based on Spice simulation. the main advantages of the proposed strategy are: full control of all steps by the student, standard language scripting, high flexibility, applicability to a wide range of analog integrated circuits, and free CAD tools.

References

[1] Group for Computer Aided Circuit Design, «Spice Opus 2.03», University of Ljubljana, Slovenia. http:\\fides.fe.uni-lj.si/spice

[2] F.L. Cox et al, «XSpice Software User's Manual», Georgia Tech Research Corp.

[3] F. Serra-Graells and N. Barniol, «Design of anlaogue Integrated Circuits: Freeware PC-based CAD for student Practices», EWME, pp.289-292, 2000

[4] F. Serra-Graells and N. Barniol, «Mixed Integrated Circuit Design with CMOS VLSI Technologies for Pre-Graduated Students», EWME, pp. 209-212, 2002

NANOWORLD PROJECTS

MONTES L., BRUGIEREGARDE L., CARDIS D., GAUBERT C., KAMARINOS G., LARTIGUE C.
Institut National Polytechnique de Grenoble (INPG), 46 Av. Félix Viallet, 38031 Grenoble Cedex 1, France

PAUTRAT J.-L., TARDIF C.
Commissariat à l'Energie Atomique (CEA), 17 rue des martyrs, 38054 Grenoble, France

CHICOINEAU L.
Centre de Culture Scientifique Technique et Industrielle de Grenoble (CCSTI), 1 place St Laurent, 38000 Grenoble, France

Opening the world of nanoscience to a large audience

1. THE PROJECTS

Micro and nano technologies (MNTs) are a real challenge and opportunity at the scientific and industrial level for the near future. They will certainly play a key role for the economy and are already the subject of many projects. In Grenoble, the INPG and the CEA, who have already initiated the Minatec Project[1], are developing different ways to communicate on nanoscience and nanotechnology. The idea is to present to a very large audience the new developments and the foreseen applications related to MNTs. It is also a way to motivate and attract students towards science education. Different approaches are being used. They are mainly oriented to high school students and teachers who can spend a day at INPG university exploring the nanoworld or, who are initiated by scientists going directly in their classroom. On the other hand, multimedia supports such as a website and a CDROM are used.
In this paper we describe these different methods and projects to make students discover and have a clear understanding of the world of MNTs, the nanoworld !

2. DOWN TO THE NANOWORLD

A website, in French for starting, "Vers le Nanomonde" (Figure 1) is under construction. It will give an overview of nanotechnologies and nanosciences and is mainly devoted to 14-18 years old students. The website is intended to be attractive, non –exhaustive, and will apprehend the nanoworld through adventures, games, music and picture galleries.

A.M. Ionescu et al. (eds.), Microelectronics Education, 175–179.
© 2004 *Kluwer Academic Publishers.*

Fig. 1: **Welcome page and picture gallery of the website 'Towards The NanoWorld**

As shown in Figure 2, the website architecture is divided in four parts. In the first part, the user is driven to the nanoworld through several adventures (a dozen is prepared by now), that are explored by four characters: two "nano-girls" and two "nano-boys".

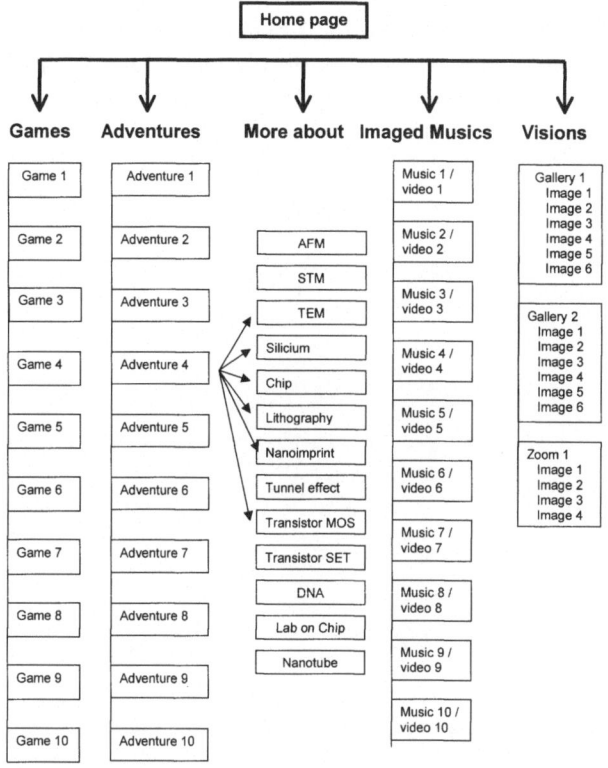

Fig. 2: **Architecture of the website 'Towards The NanoWorld'**

These adventures are focused on a specificity of the nanoworld, mainly via analogies. Each adventure is three minutes long and is punctuated by an enigma that is to be solved through a game that the net surfer has to play before getting back to the story. A simple interpretation is given, making it possible to interpret the analogy. Detailed explanations can be found in the heading "More About" containing scientific information with simple text and animations. Two other headings are revealing the beauty of the nanoworld : one mixing MNTs images and music, and one with several picture galleries (Figure 1).

3. DISCOVERING THE NANOWORLD

A second website is developed in parallel (Figure 3) dedicated to high school teachers will give them some teaching aids to approach MNTs in class. Thus the subjects are strongly correlated with the high school syllabus, even though MNTs are not yet part of it. The site comprises four headings: The first which presents general information and gives reference marks on the nanoworld, the nanometer scale, the history of nanoscience and technological projections. The other headings are very scholastics and concerns: Physics, Chemistry and Biology. As an example, carbon nanotubes are related to chemistry and are presented along with the more common carbon crystallographic structures, graphite and diamond. Biochips are linked to the study of DNA in biology, and the atomic force microscope to the study of forces in physics.

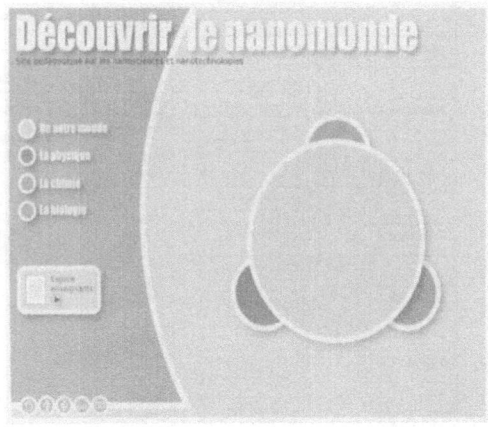

Fig. 3: **Front page of the website 'Discovering the nanoworld'**

4. A JOURNEY IN THE NANOWORLD

Practical labworks have been proposed to about 60 high school students who spent a week at university2 in december 2002 and april 2003. A full day is devoted to micro and nanoelectronics. After a brief introduction to MNTs and a tour of the different instruments used for microelectronics, students have to act : they realize a complete photolithography step in a clean room (Figure 4) using the standard silicon process facilities3. Then, it is time for discovering micro- and nanometer scales of MNTs using an electron beam microscope as well as a near field microscopes. Students observe their own fabricated structures, but also elaborated ones such as micro-electro-mechanical-systems (finger print sensors, micro-motors) that they can compare to one of their hair.

As an improvement for the next period, we will use multimedia supports such as PC-tablets to show animations describing the principle of each instrument or method. It will help in following the process flow used to fabricate a transistor.

Another way to widespread nanoscience information is done by visiting high school students in their own siteschool giving a lecture and simple labworks4.

This concept, which has been successfully applied to several domains (radioactivity, water, proteins for life, …), will be used for MNTs, explaining principle of bio-chip, "tracing DNA", or to fabricate a silicon chip: 'from sand to microchip'.

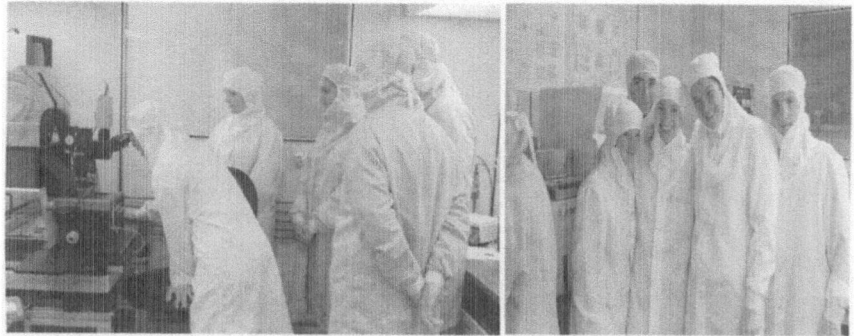

Fig. 4: In the clean room

5. CONCLUSIONS

Different projects are conducted in Grenoble with the common objective to attract high school students to science, especially MNTs. Two websites are under construction, one directly designed for students, the other one for teachers. Some students have also the opportunity to have a direct touch with MNTs through simple labworks at the University or directly in their own school. The first feedbacks of these projects are very encouraging.

We are grateful to the Région Rhône-Alpes for its financial contribution to these projects.

REFERENCES

[1]   www.mintec.com

[2]   'Classes Découvertes Ingénieur' for secondary school, INPG Grenoble.

[3]   In CIME (Centre Interuniversitaire de MicroElectronique, INPG-UJF, Grenoble)

[4]   'La Recherche fait école' - CEA

INTERNATIONAL EDUCATION

ORAL PRESENTATIONS

MARKETING AND QUALITY ASSURANCE IN RESEARCH ORIENTED MICROELECTRONICS STUDY PROGRAMS

SOARES INDRUSIAK L., GLESNER M.
Institute of Microelectronic Systems, Technische Universität Darmstadt,
Karlstr. 15 – D-64283 Darmstadt, Germany

1.   INTRODUCTION

This paper describes our experiences on introducing a new degree program in our department. Such degree is very research-oriented and thus have a narrower curricular focus than the degree programs we offered before, so we had to introduce new marketing and quality assurance mechanisms. Such mechanisms were needed to ensure that we would be able to reach a significant number of potential students for that program and to ensure the quality of the subset of them who would be admitted to the studies.

2.   INTERNATIONAL MASTER PROGRAM IN INFORMATION AND COMMUNICATION ENGINEERING

Guidelines from European Union [1] motivated the introduction of Bachelor and Master programs in the German educational system. This approach allowed the creation of degree programs - particularly in the Master level - which are specifically tailored to address particular research areas in microelectronics. Such scenario appeared as an opposition of the previous scenario, where microelectronics education was usually integrated in the curriculum of Engineering Diplom programs, thus providing only a general overview on the area. A number of advantages of the new approach can be easily pointed out:
- - graduates of such programs have a clear view of the research problems in microelectronic systems design, allowing them a quicker start in doctoral programs;
- - in many cases, graduates can already be exposed to design problems which are specific to given application domains - for instance microelectronics design for wireless communication or reconfigurable hardware for image processing – thus providing the industry with professionals with advanced practical expertise and who are able to address development challenges that are relevant to the near future.

In order to make such scenarios concrete, the study programs must have a very focused curricular structure. Following this guideline, the International Master Program in Information and Communication Engineering (iCE) was created to offer students a special blend of topics on microelectronics and communication technology. It is offered since 2001 by the Electrical Engineering and Information

A.M. Ionescu et al. (eds.), Microelectronics Education, 185–189.

Technology Department of the Darmstadt University of Technology and is tailored to both German and foreign students, as the lectures are mainly offered in English language. The program curriculum was organized in such a way that the students are prepared to deal with the convergence of computing and communication devices, such as the new generation of mobile phones or the applications of ambient intelligence. It comprehends four semesters of study. The first semester offers the mandatory foundations on microelectronics and communication technology. The second and third semesters are offered as catalogs of options, where the students can choose to increase their in-depth knowledge on more specific areas, such as system-on-chip design, communication networks and basic technologies for the realization of devices. There is also a mandatory internship, and the program is closed with the Master Thesis which is done in the fourth semester.

3. MARKETING AND QUALITY ASSURANCE

Having a narrow curricular focus, the iCE program depended on strong marketing actions to find its audience. The major challenges were:
- - reach the students which were potentially interested in such specific topic as "Information and Communication Engineering";
- - transmit to the potential students our particular view of what is "Information and Communication Engineering";
- - concentrate the marketing actions to the subset of students that would actually fulfill the program requirements.

The marketing actions we used included the use of mailing lists, website, printed posters and flyers, all built on top of the extensive network of contacts of the program faculty. We were very successful on addressing the first challenge. The response from the students was overwhelming, and as a result we received thousands of emails from potential candidates, which were followed by more than one thousand applications to join the program. As we analyzed the applications, we realized that the second and third challenges were not met, as many of the candidates seemed to have a very faint idea on the program topic and the requirements to join.

To fine-tune our marketing strategies, we decided to emphasize on the program requirements so that potential candidates would have a clearer idea about what we expect from our students. This is a particularly sensitive issue for Master programs, as the program relies on previous knowledge which is obtained by the students during their Bachelor studies. In the specific case of research-oriented master programs, the problem is maximized because the requirements can only be expressed as a set of disciplines, so there is no clear definition of a degree pre-requirement. Despite of such complicated scenario, we managed to succeed by relying on the type of material students understand the most: tests. Our first approach to better characterize our technical requirements was to include in our website some tests from our Bachelor level lectures, so the students could self-assess their background by taking those tests before applying to the program. While it improved the situation, the percentage of candidates downloading the tests was still very small. At the same time, the difficulties of our selection committee to understand and compare the different curricular structures of Bachelor-level degrees from all over the world was slowing down the selection procedure. We decided then to effectively use the tests to evaluate the background of all candidates, so that we could have a quantitative measure for quality assurance. In order to do so, we had to establish an online platform for the application procedure.

4.   ONLINE APPLICATION PROCEDURE

In order to handle the enormous volume of applications and to ensure the quality of the selected students, the whole procedure was implemented as a web-based system. The procedure starts when the candidate fills in and submits the online Application Form, available via the iCE website. The application data is then stored in a database and the candidate receives an application number. This number is a key for all the subsequent communication between the candidate and the program coordination office. After the submission of the form, the candidates have to complete the application procedure by sending via email scanned copies of the degree certificate, transcripts and English proficiency certificate (during the first application period, we accepted copies of the documents to be sent via post, fax or email, but due to the high volume of material we had to limit the submissions to digital documents only). When the scanned documents are received, a quick pre-analysis is done in order to check if the candidate's background roughly fits to the topic of the degree program. If there is a clear mismatch, the application is refused, otherwise an electronic ticket will be issued and the candidate would be allowed to take the online test.

The test is organized in several steps, each comprehends a number of multiple choice questions covering the iCE technical requirements. The questions are chosen randomly from a database, so every test is different from the others. To go from one step to the next one, a certain number of correct answers must be submit. Candidates who are not able to achieve the approval level for a given step are not allowed to access the subsequent steps. By doing so, we avoid unnecessary disclosure of the questions from our database. If a student is just taking the test to try it out, he/she will probably fail in the first steps and will not be able to access a significant number of questions. There is a time limit for each step and if such limit is not met the candidate is also not able to go to the next step. To cope with cases where the candidate's connection is not reliable, we do not require them to be online during the whole test. It is possible to download the test, disconnect, and then connect again to submit the answers as long as the time limit is not over. Figure 1 shows a sample page of the test.

The implementation of the whole application engine was done using only open source software. Figure 2 shows the major components: the Apache HTTP server, which contains a module to interpret PHP scripts, and through PHP it is able to connect to MySQL, a relational database.

In order to validate the online test, a presential test covering exactly the same topics is done when the students arrive in the university, thus ensuring that the students doing the online test are actually the same that were admitted.

CONCLUSIONS

Our experiences have shown us that the importance of marketing strategies and quality assurance is increasing as more research-oriented study programs are introduced. While such programs must rely on mass marketing to create awareness among potential candidates, the quality assurance mechanisms are even more important to make sure that the right candidates are the ones to be admitted to the studies. Instead of relying on intelligence tests or those covering wide ranges of engineering disciplines, we implemented an online application procedure that aims to a win-win situation where the students know what is expected from them in this specific degree program, and at the same time the university ensures the quality of the alumni.

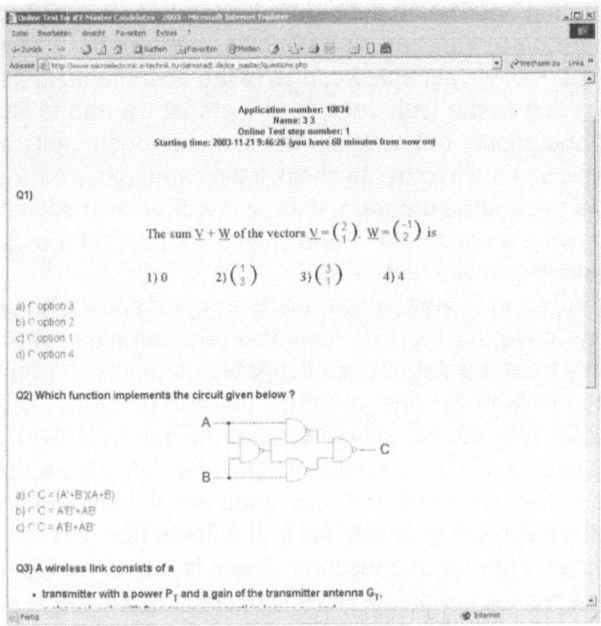

Fig. 1: iCE online test sample page

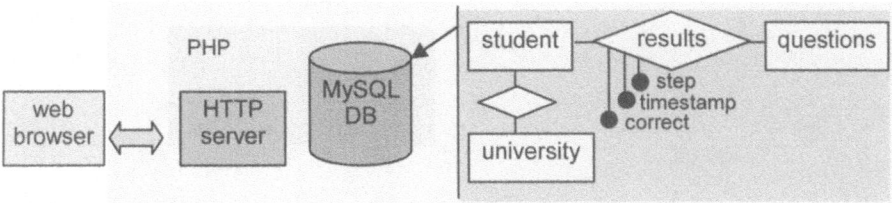

Fig. 2: Technical infrastructure of the iCE online application engine.

REFERENCES

[1]   The European Higher Education Area. The Bologna Declaration of 19 June 1999.

SHARING ONLINE LABORATORIES AND THEIR COMPONENTS

A new learning experience
JEPPSON K.O., LUNDGREN P., DEL ALAMO J.A., HARDISON J.L.*, ZYCH D.*
*Chalmers University of Technology, MC2, SE-41296 Göteborg, Sweden, *Massachussetts Institute of Technology, Cambridge, MA, USA*

1.   INTRODUCTION

In this paper we report on the use of the online MIT Weblab system [1] for characterization of semiconductor devices in three qualitatively rather diverse microelectronic device courses offered by Chalmers University of Technology, including junior undergraduate courses as well as extension courses. In particular we will focus on the learning situation and the impact of class size. Since the laboratory equipment is available online 24 hours-a-day every day during the course, new opportunities for integrating laboratories into the learning process have become available. In particular, we will discuss the role of assignment formulation to support this new learning situation.

In this paper we will describe our experiences from using the MIT online laboratory to shift student focus from instrument handling to data analysis, parameter extraction, and model fitting. This can be done through rather open lab assignments where the students themselves can organize the details of their specific task within the context of the overall objective of the laboratory exercise.

2.   About weblab

In a topic like microelectronic device physics the student learning experience can be substantially enhanced by hands-on characterization of diodes and transistors. However, for a variety of practical and economic reasons universities have found it more and more difficult to include such a laboratory component. A remote laboratory available over the internet solves many of these concerns while largely preserving, or even enhancing, the educational experience. Online remote laboratories not only offer the possibility to perform traditional laboratory exercises in a more cost effective way, but they also make available to students more advanced instruments than have traditionally been affordable. Many institutions in different fields have explored this concept of an online laboratory. One such joint European remote laboratory network is presently being developed within the EU Socrates/ Minerva framework [2].

Over the last few years, MIT has been experimenting with a system called the MIT Microelectronics WebLab. This system allows microelectronic device characterization through the world wide web. Through WebLab, students can take current-voltage measurements on transistors and other microelectronics devices in real

A.M. Ionescu et al. (eds.), Microelectronics Education, 191–195.

time from anywhere and at any time. The basic architecture of the system and its use in a variety of educational settings was reported in [3].

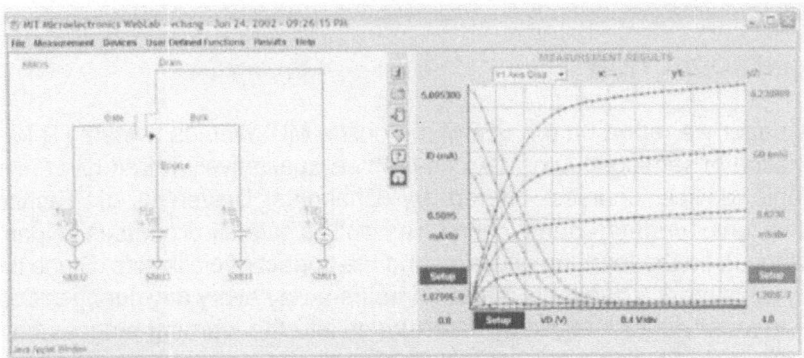

Fig. 1: Screen-shot of WebLab graphical interface: the main window

The user interface for WebLab is a Java applet which duplicates the essential functionality of the analyzer's console, see Fig. 1, allowing the user to set up a measurement for one of the devices that is currently connected to the system (the necessary information about these devices is provided by the server when the applet loads). When the user is ready to execute a measurement, the applet sends the measurement specifications to the server. More details of the WebLab system and its graphical interface are given elsewhere [4].

3. Short Description of Chalmers courses and Mission task

The WebLab has so far been used remotely in three different courses offered by Chalmers University. Following two small test runs (one in an elective graduate course (eight users) and one in a extension course offered to professionals working in local industry (six users), WebLab was employed in a large compulsory junior undergraduate course with about 330 students during the spring of 2003.
In all courses students were given a clear objective of the laboratory task and what was expected of them. A simple instruction was given that advanced technology transistors of four different types were available through MIT WebLab.
Examination of the lab assignment in the undergraduate course was performed through group meetings where an examiner directed individual questions to the lab group members who were to respond with the help of a whiteboard. Individual credits were rewarded to the group members according to performance in this oral examination. The communication between MIT WebLab administration and Chalmers course management was conducted by e-mail and for the two small courses the planning could be settled with some ten mails and replies.

4.   Evaluation

The online laboratory experiments were evaluated through detailed discussions with students in the graduate courses and through written review questionnaires handed in by students in the undergraduate course. The overall impression on the use of online laboratories among engineering program students was generally very positive. A summary of the evaluation regarding system access and stability, user friendliness, and educational value is shown in Fig. 2.

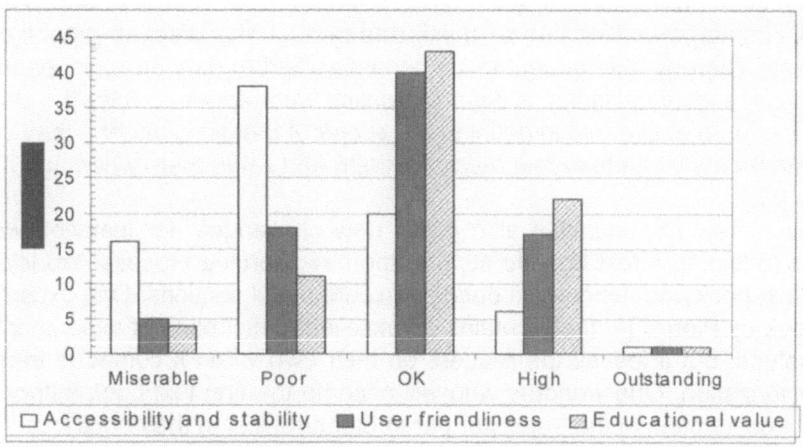

Fig. 2: Summary outcome of evaluation questionnaire

Students appreciated most that they could decide themselves when to do the laboratory exercises. This was perceived as "less stressful" [than traditional eve-ning laboratory classes]. They also appreciated the graphical interface, because "you could see the graphs clearly" and it was "excellent when analysing data". The most severe drawbacks were the system instabilities. The system was instable when many students were logged on simultaneously: "First you could not access the home-page at all, and then "once you managed to log in it kept on crashing".

5.   Outcome

The critical difference of a large student class
The use of WebLab in the undergraduate course at Chalmers was the largest and most ambitious deployment of WebLab to date. On Feb 25, 2003, there were 134 characterization experiments executed in a single hour (on average that means a job every 27 seconds). An experiment this scale was bound to result in the iden-tification of new problems that had never been seen before when operating at lighter loads. Two kinds of problems were encountered. First, there was a handful

of system blackouts during which WebLab was unavailable for measurements. Second, the system returned an error message in response to a valid experimental setup. The origin of both types of errors was identified and corrected.

DISCUSSION and conclusion

In our experiment with online laboratories we had an explicit purpose to get away from traditional closed-form laboratories. In that type of laboratory students devote most of their time to handling of the instruments to collect device data following a step-by-step instruction manual, frequently asking the teaching assistant for help to find a short-cut to the next step. In an online computer-based laboratory, instrument handling can be minimized through the WebLab graphical user interface. Thereby, the student focus can be shifted to data analysis, parameter extraction, and model fitting. In essence, online laboratories enables the students to take a more active role in defining the scope of the assignment – they can do measurements when they feel ready for them and re-do them when and if need arises [5].

However, new opportunities also mean new challenges, for instance when it comes to finding a text book to support the new learning process. Traditionally, most text books do not contain detailed experimental sections. One exception is the book by Pierret [6] that contains an excellent description of experimental diode setups, but it leaves the readers on their own when it comes to transistor characterization. One wonders who will become the first text book author to include a description of the transistor parameter analyzer in their book?

Some problems encountered in the course of this experiment had a more negative impact on the overall experience of the undergraduate students at Chalmers University when compared to MIT students using WebLab in MIT courses. There are two reasons for this. First, at Chalmers students worked in groups, while at MIT students assignments were of an individual nature. This is relevant because at Chalmers, students had to make an appointment to work together on their lab assignments at a specific time. If the system was not available or if the system did not operate properly at that very time, students were forced to reschedule leading to frustration and possible project delay. In an individual assignment, a student has a lot more flexibility to schedule their work and the consequences of system malfunction are much less severe.

The second reason for the negative impact of systems problems in the overall educational experience of the Chalmer's students is the time zone difference between Chalmers and MIT. As a consequence, several hours could pass be-tween the occurrence of a system problem and its satisfactory resolution, this even if the problem was of a trivial nature and its solution would only take a few seconds.

REFERENCES

[1]   J. A. del Alamo, J. A., L. Brooks, C. McLean, J. Hardison, G. Mishuris, V. Chang and L. Hui, "The MIT Microelectronics WebLab: a Web-Enabled

Remote Laboratory for Microelectronic Device Characterization", World Congress on Networked Learning, Berlin (DE), 2002

[2]   R. Cabello et al, "eMerge: An European educational network for dissemination of online laboratory experiments", ICEE, Valencia (ES), 2003

[3]   J. Henry, "Running Laboratory Experiments via the World Wide Web", ASEE Conference, 1996

[4]   J. A. del Alamo, V. Chang, J. Hardison, D. Zych, and L. Hui, "An Online Microelectronics Device Characterization Laboratory with a Circuit-like User Interface", ICEE, Valencia (ES), 2003

[5]   A. Söderlund, K. O. Jeppson, F Ingvarson, and P Lundgren, "The Remote Laboratory – A New Complement in Engineering Education", ICEE 2002, Manchester (UK), 2002.

[6]   R. F. Pierret, "Semiconductor Device Fundamentals", Addison-Wesley, 1996

EDUCATIONAL AND EUROTRAINING CURRICULUM IN DATA ACQUISITION AND SIGNAL PROCESSING FOR SMART SENSORS

YURISH S.Y.
International Frequency Sensor Association (IFSA), Bandera srt.,12, Lviv, 79013 Ukraine, http://www.sensorsportal.com, e-mail: info@sensorsportal.com

1.   INTRODUCTION

Smart sensors are of great interest in many fields of industry, control systems, biomedical applications, etc. It is very important research and educational area for European economical and technological bases. As rule, the majority of training courses over the past years dedicated to smart sensors are reflecting only technological achievements of microelectronics and materials [1-3]. Although a number of universities have courses in sensor technology – specifically in silicon-based and micro-machining technology [4], such technology oriented educational curriculum give deep knowledge "how to make", but not "what to make". However, modern advanced microsensor technologies require novel advanced measuring and data acquisition technique.

The main task of measuring instruments, sensors and transducers designing has been always to reach high metrology performances. At different stages of measurement technology development, this task was solved by different ways. They were technological methods, consisting in technology perfection, as well as structural and structural-algorithmic methods. Historically, technological methods have received prevalence in the USA, Japan and Western Europe. The structural and structural-algorithmic methods have received a broad development in the former USSR and go on developing in NIS countries now. The improvement of metrology performances and extension of functional capabilities are being achieved through implementation of particular structures designed in most cases in heuristic way using advanced calculations and signal processing. Smart sensors and transducers are not the exception [5].

As an initial step to fulfill this urgent need, International Frequency Sensor Association (IFSA) has initiated a new educational program in data acquisition and signal processing for smart sensors [6]. Utilizing the existing many years research expertise, one semester course as well as the seminar have been created and tested.

2.   CURRICULUM DEVELOPMENT

There is a need to integrate state-of-the-art research on smart sensors and novel conversion methods into the curriculum both in terms of formal lecture presentations, PC demonstrations and practical tasks solution.

A.M. Ionescu et al. (eds.), Microelectronics Education, 197–201.
© 2004 *Kluwer Academic Publishers.*

The devices covered in this course include industrial state-of-art smart sensors as well as novel intelligent self-adapting sensors and transducers. The course materials are developed from the current and past research IFSA's projects. The prerequisites for this course are the following: physics, measuring technique, uncertainty analyses, semiconductor devices and technology, VLSI design. In addition to these, the students will also take required standard courses from the traditional programme in respective departments to satisfy their degree requirements and work on original research projects as part of their thesis or dissertation.

3.   COURSE CONTENTS

This new course sequence is designed to fulfill the demand of high-technology industries and research laboratories for students with a strong background and hands-on experience in advanced electronic, signal processing and computer simulation techniques. The course sequence topics are outlined below.

Introduction.
- 1) Smart Sensors for Electrical and Non-Electrical, Physical and Chemical Variables: Tendencies and Perspectives.
- 2) Data Acquisition Methods for Multichannel Sensor Systems.
- 3) Methods of Frequency-to-Code Conversion for Smart Sensors.
- 4) Advanced and Self-Adapting Methods of Frequency-to-Code Conversion.
- 5) Signal Processing in Quasi-Digital Smart Sensors.
- 6) Digital Output Smart Sensors with Software-Controlled Performances.
- 7) Multichannel Intelligent and Virtual Sensor Systems.
- 8) Software Level Smart Sensors Design.
- 9) Smart Sensors Buses and Interface Circuits.
- Summary and Future Directions.

Recommended references for the further reading are the following [4-7].

4.   SENSOR WEB PORTAL'S CONTENT, HAVING EDUCATIONAL IMPACT

The current and likely impact of the Internet in the world is in education, business and researches. An education on the nature is one of most information consumption processes; therefore, the progress in an education directly depends on development of information technologies. In new ways of e-teaching and e-learning conditions, information content plays significant role for virtual education.

The developed curriculum is strongly connected with high technological content and scientific basis of specialized (vertical) monthly up-dated Sensors Web Portal

[6]. Its content is strongly oriented to research and education in smart sensors, measurements, data acquisition and microsystems (MEMS) high technological areas. It is the technically most sophisticated and most flexible dissemination information service tool for distribution of current research information on sensor technology based on the World Wide Web (Internet), which has been initiated and is funded by International Frequency Sensor Association (IFSA).

The main aim of launched in 1999 sensors web portal is to provide a forum for academicians, researchers, Ph.D. students and engineers from industry to present, discuss and study the latest research results, experiences and future trends in the area of design and application of different modern sensors. Sensors Web Portal is organized like sensor resources directory, aggregating information from various sources and presenting it in a user-friendly format.

The Training Course section (Figure 1) focuses on sensors related training courses for students, post graduate students and professionals (researchers and engineers). It showcases some of the more recent calls for EuroTraining courses as well as new university courses. There are some free on-line courses, for example, 'Basis of sensors course', etc.

The Sensors section divided into the following subsections: Acoustic, Biosensors, Chemical, Flow, Gas, Humidity, Magnetic, Mechanical, Optical, Pressure, Proximity, Resonant, Rotation speed, Temperature, Tilt, Torque, Ultrasonic, Vacuum and Others subsections, including Sensor Instrumentation (Figure 2)

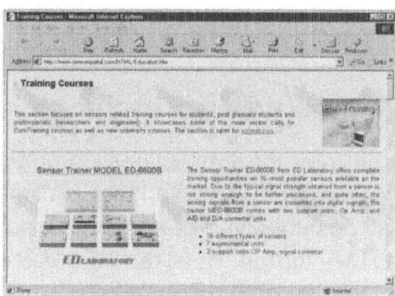

Fig. 1: Training courses web page

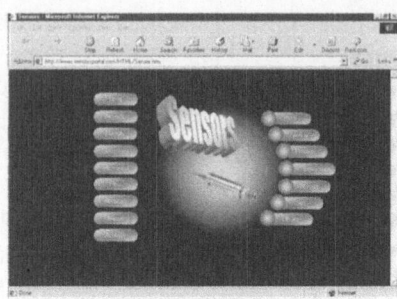

Fig. 2: Sensors web page

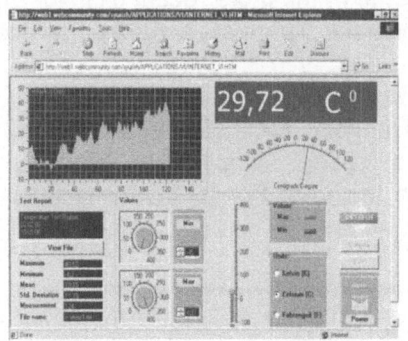

Fig. 3: nternet application - virtual thermomete

There is a possibility to run Internet applications demonstrating the virtual measuring instruments on-line (Figure 3). The Sensors section is quickly growing and will include in the future all possible type of sensors. Traditionally, the majority of the information is in the HTML format. Papers and articles (in pdf – and doc- formats) are compressed by WinZip archive software.

The executable files include the Internet applications and Java applets. The classified by certain way information, placed on Sensors Web Portal, allows receiving the information about state of the art and researches in the area of modern sensors. The user, in turn, can select the particular type of sensor as well as the modern method of frequency-time measurements for the usage or studying.

Creating educational web sites, e-learning systems and virtual universities it is not necessity now to create the content again for sensors related training courses. With the aim to save time and money it is expediently to give appropriate links to specialized vertical portals, for example, to Sensors Web Portal and use its huge content at different training course creation as well as its stored articles, papers, books, patents, abstracts, etc. like e-libraries for students and researches.

5.   CONCLUSIONS AND RESULTS

During 2001-2003, the developed course has been given in European universities in Germany (University of Applied Sciences, Furtwangen, twice), Italy (Technical University of Perugia), Spain (Technical University of Catalonia in Barcelona) and Austria (Graz University of Technology). The lectures from this course has been included into annual quality labeled EUROPRACTICE course on 'Smart Sensor Systems' in 2001-2004, Delft University of Technology (The Netherlands) and NATO Advanced Study Institute 'Smart Sensors and MEMS' [8].
The further development of the proposed course - Data Acquisition and Signal Processing in Smart Sensors and MEMS - has been included into IEEE Sensors Council on-line tutorial program (high quality tutorial-type lectures from sensors experts) and will be made available on-line through the IEEE Sensors Council web site [9]. Due to these efforts better trained researchers and engineers will be competing in tomorrow's job market.

REFERENCES

[1] G.W. Auner, P. Siy, R.Naik, L.Wenger, G-Y. Liu, L.J. Schwiebert "A combined research/educational curriculum in smart sensors and integrated devices", *In Proceedings of International Conference on Engineering Education (ICEE'98)*, Rio Othen Palace, Rio de Janeiro, Brazil, 17-20 August, 1998.

[2] S.A. Akbar, P.K. Dutta, M.J. Madou "Novel sensors R&D leading to curriculum development", *In Proceedings of International Conference on Engineering Education (ICEE'98)*, Rio Othen Palace, Rio de Janeiro, Brazil, 17-20 August, 1998.

[3] S.A. Akbar, P.K. Dutta, "A research driven multidisciplinary curriculum in sensor materials", *http://www.mse.eng.ohio-state.edu/~akbar/asee.htm*

[4] S.Middelhoek, S.A.Audet, P.J.French, "Silicon sensors", *Lecture ET 4-046*, Delft University of Technology, Delft, The Netherlands, 2000.

[5] N.V. Kirianaki, S.Y.Yurish, N.O.Shpak, V.P.Deynega "Data acquisition and signal processing for smart sensors", *John Wiley & Sons*, 2002.

[6] Sensors Web Portal, *http://www.sensorsportal.com*

[7] *Smart Sensor Systems, EUROPRACTICE Advanced Engineering Course Materials*, Delft University, The Netherlands, 2003.

[8] *Smart Sensors and MEMS*: Tutorials and Posters Abstracts, NATO ASI, Preprints, Ed. by Maria Teresa Gomes and Sergey Y. Yurish, Povoa De Varzim, 8-19 September 2003 (ISBN 0-9733840-0-X).

[9] IEEE Sensors Council web site: http://www.ieee.org/sensors

MICROSYSTEMS

ORAL PRESENTATIONS

SETTING A TUTORIAL ON THERMAL POLYSILICON-BASED SUSPENDED ACTUATOR USING SURFACE MICROTECHNOLOGY

SALAÜN A-C., KOTB H-E.M., COULON N., ROGEL R., MOHAMMED-BRAHIM T., BONNAUD O.
Groupe Microélectronique, IETR, pôle de Rennes du CNFM : CCMO, Université de Rennes I, Campus de Beaulieu, 35042 Rennes Cedex, France.
E-mail: anne-claire.salaun@univ-rennes1.fr, Olivier.bonnaud@univ-rennes1.fr

Abstract

This paper deals with a tutorial for students at master level that consists in the realization of moving microstructures based on polysilicon thin films. The final goal is to fabricate a generic moving suspended structure based on thermal expansion difference of two arms, which can be involved as micro-switch or micro-engine in a more complex micro-mechanical system. Optical and electrical characterizations are done during the process steps and the structures are checked at the end of the fabrication process using, on one hand scanning electron microscope (SEM) and on the other hand, under electrical excitation, optical microscope. After this tutorial, students are familiarized with the different aspects of microtechnology and micro-mechanisms, in particular, with technological exigencies, thermal and stress problems, and on the role of the scale factor.

1.   Introduction

Surface micromachining represents an important tool box for Micro-Electro-Mechanical Systems (MEMS). The great advantage of this technology is its compatibility with the conventional IC techniques, which makes possible the integration of MEMS devices with the electronic driving components. For these reasons, we have decided to familiarize our students with the basics of surface micromachining. A first step was set up by fabricating basic elements of the surface microtechnology, i.e. cantilevers, bridges and membranes involving either metals or polycrystalline silicon [1]. In these techniques, device components were fabricated on the surface of different substrates through a series of deposition, lithography and selective removal of thin films.

To fabricate free-standing surface micro-machined structures, it is necessary, first to deposit hard film on sacrificial layer that can be removed, and second to reduce the stress in the structural layer to a minimum value. Compressive stress can result in film buckling, while high tensile stress may lead to film cracking. Many methods are currently used to measure stress [2-4]. Among these methods, we have chosen to measure stress using surface micro-machined cantilevers, bridges and micro-rotating structures as presented in previous works [1]. These test cells are thus included in the design of dedicated structures.

A.M. Ionescu et al. (eds.), Microelectronics Education, 207–212.

The microstructures are fabricated using sacrificial and structural layers. In our case and on the basis of laboratory experience, the structural films are polysilicon made that can be *in-situ* doped in order to be conductive. Polysilicon, usually used in MEMS, exhibits, at the same time, a relatively good electrical quality, the heavily doped films are not too much resistive after crystallization, and also acceptable mechanical properties in term of hardness, residual stress, and stress gradient. The sacrificial layer can be a deposited oxide or, as developed in the research team of our Common Center of Microelectronics, deposited germanium. This last material gives the advantage to be easily and fast etched in de-ionized water (DI) with a very good germanium/polysilicon selectivity [5-6].

2.   Educational objective

We propose a technological student project in the cleanroom of the Centre Commun de Microélectronique de l'Ouest (CCMO), pole of the French Committee for Education in Microelectronics (CNFM) [7]. This tutorial, which corresponds to six full days of process in the cleanroom, is proposed to master degree students – yet called DESS students (bac +5 level) - in their long duration laboratory project, with good prerequisites in microelectronics technology. The depositions of polysilicon and poly-germanium using Low Pressure Chemical Vapor Deposition techniques are made with the help of technicians and permanent staff due to the dangerousness of silane and germane gases involved in these steps. Electrical and optical characterizations are performed during the tutorial and, at the end of the process, microstructures are observed under electrical excitation.

3.   Processing steps of surface micromachining

The masks were initially designed to get several lengths and several widths basic structures as well some solid membranes as lighted membranes (holes inside), in order to detect the effect of the size and of the geometry on the basic structures and on final free-standing membranes. The presence of holes in the membrane also allows improving the sacrificial film etching located under the structural one. Figure 1 describes the main process steps of surface micromachining fabrication on silicon wafers.

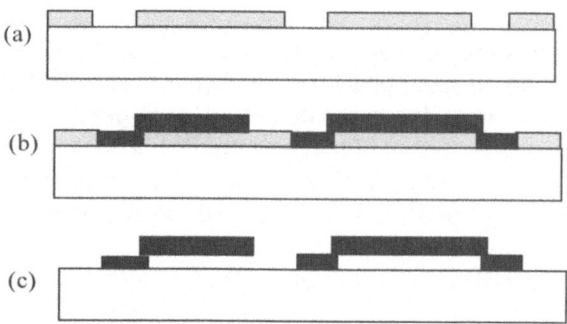

Fig. 1: Schematic description of surface micromachining : a) sacrificial layer is deposited and patterned, b) structural layer is deposited and patterned and c) sacricial layer is removed and structures are released. The left hand side structure can be cantilever or released membrane

The sacrificial layer is a 0.5μm thick germanium layer deposited by Low Pressure Chemical Vapor Deposition (LPCVD) on silicon substrate. Let us note that they can be also deposited on oxidized silicon substrate as well on glass substrates, the proposed technology being low temperature and large area oriented. After patterning, 0.5μm thick phosphorus *in-situ* doped polysilicon layer is deposited also by LPCVD technique and patterned. Polysilicon film is patterned and plasma etched (SF_6 Reactive Ion Etching) to define the different structures. Finally, the samples are immersed in DI water and hydrogen peroxide solution to release the suspended structures. Samples are rinsed in DI water then ethanol to reduce the surface tension forces. This step avoids the stiction of the suspended structures. Finally, substrates are dried inside an oven at 120°C.

4.   Challenges

In surface micromachining, we have to:
- choose a structural material hard enough to resist to mechanical constraints
- minimize the stress and stress gradient in the structural layer,
- find a sacrificial layer which can be etched very selectively against the structural layer,
- avoid the stiction of the released microstructures

In this case, we have in addition to create an excitation high enough to generate a moving of the membrane.

4.1   CHOICE OF THE SACRIFICIAL FILM

Following the technique well controlled in the laboratory and resulting of specific studies in fabricating air-gap polysilicon thin film transistor [5-6], the chosen sacrificial layer is germanium deposited by LPCVD with germane as precursor gas. This deposition is performed at 400°C and the film is polycrystalline (as-deposited) in this condition. The etching rate can reach 1μm/minute in a solution of hydrogen peroxide (H_2O_2) at 90°C with a very high selectivity (more than 1000).

4.2   CHOICE OF THE POLYSILICON STRUCTURAL LAYER

The silicon layer is deposited in amorphous state using LPCVD technique at 550°C and then solid phase crystallized at 600°C. Polysilicon layers are *in-situ* doped by incorporation of doping atoms, phosphorus (type N) in our case. This choice is lead by the fact that phosphorus doped film exhibits a slightly tensile mechanical stress gradient at least much lower than in the case of boron doped one. This avoids any bending of the suspended structure after releasing.

5.   Final fabricated structure and characterization

On the same substrates than the cell tests, released structure actuators were fabricated. Figure 2 shows a photograph of the structure layout. The principle of this actuator is the difference of thermal expansion of the both arms of the membrane induced by the difference of their own temperature.

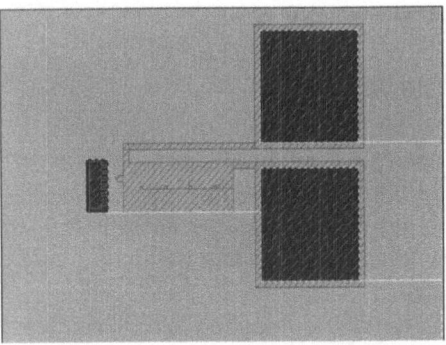

Fig. 2: mask layout of the released membrane. The displacement is produced by the difference of heating in the narrow arm compared to the wide one. The two pads are used to put the biasing probes. The difference of the thermal expansion of the both arms induces a moving of the membrane. This moving is visualized thanks to an index

This difference of temperature is generated by the difference of electrical resistance of the both arms that are submitted to a current flowing from the pads. The difference of electrical resistance comes from the difference of the conduction section. The power dissipation generates an increase of temperature in the both arms, in a short enough time.

Figure 3 shows a photograph of a final structure corresponding to the layout shown figure 2. The electrical characterization is performed in a electrical probing station under optical microscope. The presence of the index allows the visualization of the moving of the membrane when the structure is biased. The wide arm contains several holes that increase the etching rate of the sacrificial layer. Let us note that a short video could be projected during the conference sessions.

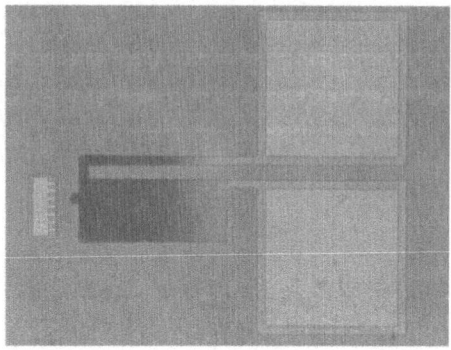

Fig. 3: thermal polysilicon based actuator. The index allows controlling the deviation of the membrane when the structure is biased. The holes in the wide arm (lower on the figure) improve the etching of the germanium sacrificial film.

Conclusions

The setting of the micro-mechanical tutorial allows familiarizing graduate students to microtechnology processes in a cleanroom including several types of problems, such as deposition of polycrystalline materials in a low temperature technology, etching of sacrificial layer under structural film, control of the mechanical stress, and characterization of microstructures. This pedagogical experience allows processing some micromechanical generic structures with graduate students. Then, this tutorial constitutes a good experience for the students that also enjoyed a lot this original work.

References

[1]   A.C. Salaün, H-E. Kotb, T. Mohammed-Brahim, F. Le Bihan, N. Coulon, H. Lhermite, O. Bonnaud, Setting of microtechnology tutorials using surface micromachining, EWME'2002, Poster, Vigo (Spain), May 2002, Proceedings of 4th European Workshop on Microelectronics Education., pp. 145-148.

[2]   M.S. Benrakkad, M.A. Benitez, J. Esteve, J-M. Lopez-Villegas, J. Samitier et J.R. Morante. J.Micromech.Microeng. 5, 132-135, 1995.

[3]   X. Zhang, T.-Y. Zhang et Y. Zohar. Thin Solid Films 335, 97-105, 1998.

[4]   J.Singh, S.Chandra et A.Chand. Sensors and Actuators 77, 133-138, 1999.

[5]   H. Mahfoz-Kotb, A.C. Salaün, T. Mohammed-Brahim, N. Coulon, O. Bonnaud and J.Y. Mevellec, High performance polysilicon air-gap thin film transistor on low temperature substrates, Proc. SPIE, Smart Sensors, Actuators, and MEMS, Vol. **5116**, (2003), pp.168-175

[6]   H. Mahfoz-Kotb, T. Mohammed-Brahim, A.C. Salaün, F. Le Bihan, Polycrystalline Silicon Thin Films for MEMS Applications, Thin Solid Film,s Vol. **427**, pp.422-426 (2003)

[7]   O. Bonnaud, G. Rey, "The French microelectronics training network supported by industry and education ministries", *Proceedings IEEE Int. Conf. on Microelectronic Systems Education, (MSE'97)*, Arlington-VA (USA), July 1997, pp. 121-122.

TECHNOLOGY OF ELECTRONIC DEVICES – A SINGLE COURSE

ISAI G., MOUTHAAN T.
University of Twente, Electrical Engineering Department, P.O. Box 217, 7500 AE Enschede, The Netherlands

1. INTRODUCTION

Learning how to fabricate and design various microsystems is very important for the future electrical engineers. Technology courses present not only knowledge about the existing techniques but offer also the necessary tools for development of future devices.

Usually, each group from the department of Electrical Engineering has a technology course which is adapted to its specific area of research. For example, the MEMS group provides the course: MEMS based technology; the semiconductor group presents IC technology, etc. The courses are not identical; however, the fabricating technology is similar for very different devices such as: sensors, actuators, MRAM, MOSFET transistor, biomedical sensors, LED, etc. In order to make such microsystems, one uses the same basic processes: film deposition, modifying the film properties, making a pattern in the film with the help of lithography and etching, planarisation, packaging, etc. The idea of introducing a single technology course is therefore the next logical step. Instead of increasing specialization, the course will address technology as a whole, preparing therefore engineers which are more adaptable to the needs and the requirements in industry.

Because microtechnology and nanotechnology are related and it is important to train the students for the future, the course should contain also the limits of microtechnology, the latest innovations and especially information about nanotechnology.

2. DIFFERENCES BETWEEN TECHNOLOGY COURSES

The differences between the technology courses result from the fact that different groups use different properties of the materials; the accent is put either on electrical properties, on magnetical properties, on optical properties, or on mechanical properties of the films. These differences are however not major, considering that the fabrication techniques and equipments are mostly the same and just the processes of optimization the desired property are different. By comparison a material science course is much more heterogeneous than a general technology course would be.

A.M. Ionescu et al. (eds.), Microelectronics Education, 213–216.

3.   THE ADVANTAGES OF HAVING A SINGLE COURSE OF TECHNOLOGY

At the University of Twente, after the bachelor program of three years, the students can extend their knowledge by entering the master studies of two years. One of the master programs in Electrical Engineering department is "Microsystems and Microelectronics" [1], in which the five groups shown in figure 1 are active. Teaching each year three technology courses (figure 1) which have in common around 70% of the content is a waste of resources. For more efficiency, a microtechnology course for all the students will replace the three courses which were given within the groups. The course will be part of the compulsory courses given at the beginning of the master program. It is an introductory course with fundamental knowledge and will prepare the students for fabricating various types of devices. After the compulsory courses, the students will choose a group to continue their studies and learn more about the functioning and applications of microsystems in specialized courses.

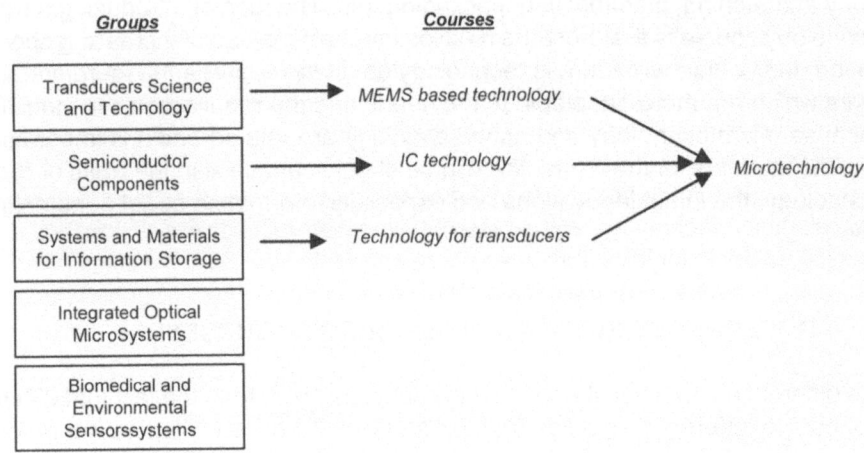

Fig. 1: Technology courses in the Microsystems and Microelectronics master program

There are also important educational reasons to transform the specific technology courses into a single course. In a general technology course it can be shown how different types of devices can be integrated on the same chip. For example integrating MEMS and ICs on the same chip can be done either by Berkley or Sandia approach. This cannot be shown in the specialized technology courses. Furthermore, an interesting subject which can be given as a homework would be to imagine an integrative microsystem which includes a few sensors (optical, mechanical), actuators, an IC, a memory, an interface with biological systems, all on the same chip. This integrated system, while not yet realizable in practice, is a future goal for the industry, due to its huge number of applications. By challenging and stimulating the students with this new perspective on technology in which the focus is on technology flows for integrated devices, it is possible to prepare them for the future.

Another advantage of the single technology course would be the fact that the student will be able to observe the multitude of devices and applications that can be realized with the help of technology. Their knowledge and skills would be also broadened, considering the fact that the course contains general information useful in more than one area. Therefore the students will be better prepared for solving complex problems, will have more flexibility in choosing a job and will adapt better to the mobility of the industry and of the job market.

4.   TECHNOLOGY COURSE ORGANIZATION

The content of the course can be organized as:
1. Introduction
2. Substrate selection
3. Film deposition and growth
4. Modifying the film structure/properties
5. Defining patterns
6. Materials and process characterization
7. Integration of process steps to build devices
8. Future (nanotechnology)

The lack of books on general technology is the driving force in conceiving the material for this course. The accent in the technology course should be put on the description of the phenomena, techniques and equipments involved in growing and modifying a film and defining a structure in the film. Emphasize should be put also in indicating the elements that are not common in fabricating different types of devices. For example while silicon is the dominant materials in fabricating ICs, GaAs is used mostly in optoelectronic devices. Fabrication, modification and characterization of both materials should be treated in the course.

The students should understand at the end of the course the possibilities, but also the limits of technology. Furthermore, they should learn how to choose the appropriate system for fabricating or designing a specific component. This can be realized by comparing different techniques and showing the advantages and disadvantages of each. The students should also be able to observe how the same equipment can be used to make different layers with very different properties. For example the plasma can be tuned to deposit slowly high-quality thin gate dielectrics or to obtain quickly thick dielectrics with poorer quality. The manner in which the parameters of the equipment influence the film properties can be understood by knowing the physical processes that take place. The science behind the device fabrication is more important than the details related to techniques, conditions, etc.

Nanotechnology should be a significant part in the course. It is important for the student to have an image of the development of technology (the history in the introduction, the present during the course and the future at the end of the course). The student should understand the limits of present technology, should be aware of the latest developments and should have an idea about how the future will look like. Furthermore, the present student is the future researcher; therefore it is useful for him to have some basic knowledge about the nanoscience and nanotechnology. Due to the miniaturization process and to the numerous applications, nanotechnology is a "certainty" in the future. Although nanodevices do not exist at this moment, a brief exploration of the possible routes toward the unknown can be not only useful but also very attractive and challenging for students.

Another important concept is decreasing the barriers between theory and problems. The student should be permanently bombarded by questions from the teacher during the lectures. Furthermore the solutions for problems should have references to theory. In this manner the accent is put on problem-solving which will be helpful for the future electrical engineers.

CONCLUSIONS

The technology of fabricating microsystems can be included together with the material science in fundamental knowledge for the electrical engineering students. The lack of general technology courses is a result of specialization and departments' separation into groups. Teaching multiple technology courses is a waste of university resources. Not only the university but also the students would beneficiate from a single course of technology. The students would be able to understand the multitude of possibilities and applications that can be realized with the same fundamental process steps. They would be able to address the fundamental issues in technology instead of learning only about fabricating a particular device. Furthermore they would acquire a broader knowledge and flexible expertise.

REFERENCES

[1]   A.J. Mouthaan, "A case study of a microsystems curriculum", COMS2003, Amsterdam, The Netherlands, September 2003.

STUDY OF A MICROWAVE COUPLED LINES FILTER: FROM CLEAN ROOM REALIZATION TO MEASUREMENT AND BACK-SIMULATION.

LISSORGUES G.
ESIEE, Cité Descartes, BP99, 93162 Noisy-Le-Grand, FRANCE.

1.   Context of the study

In direct connexion with the theoretical approach of RF and microwave passive filters, this experimentation was developed with three main goals:
- discover some clean room technological steps through the realization of their own circuit on alumina
- discover the network vector analyser measurement set-up dedicated to microstrips
- link theory of coupled lines filters with experimental results, in particular in term of sensitivity on the technological process, using ADS back-simulation possibilities.

This study may be extended to other microwave circuits on alumina, such as RF mixer using PIN diodes. It would show to the students bonding requirements in addition to the microstrip technology. But the idea was first to illustrate the course on RF and microwave passive filters. Due to the wide variety of filters and associated technologies (distributed, coupled lines, waveguides, SAW...), it was important to complete the course with an example.

2.   DESCRIPTION OF THE FILTER

2.1   GEOMETRY OF THE FILTER

The selected topology was a parallel type coupled lines band-pass filter centred at 12.3GHz on a 2" alumina substrate with ε_r = 9.8 and height H = 635µm, covered with 12µm thickness of gold metallizations. We used ADS correlated with Momentum software to define the dimensions, as shown on figure 1. The extracted layout is presented on figure 2.

2.2   FIRST SIMULATIONS USING ADS TOOLS

During a previous tutorial on different kinds of filter technologies with the same students, we verified the transmission (S21) and reflection (S11) behaviour of the filter, as shown on figure 3. Students have also been introduced in this tutorial to tuning techniques, as well as optimisation and yield tools, to be able to perform the back-simulation required after the measurement step.

A.M. Ionescu et al. (eds.), Microelectronics Education, 217–221.

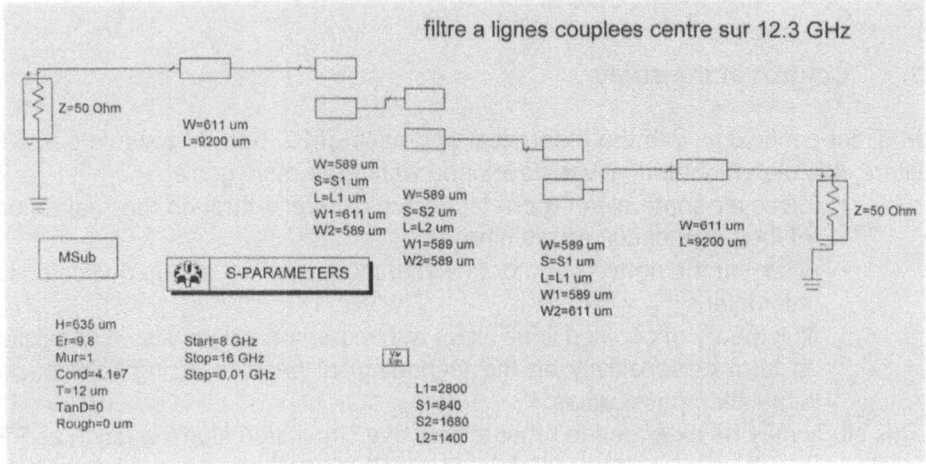

filtre a lignes couplees centre sur 12.3 GHz

Z=50 Ohm

W=611 um
L=9200 um

W=589 um
S=S1 um
L=L1 um
W1=611 um
W2=589 um

W=589 um
S=S2 um
L=L2 um
W1=589 um
W2=589 um

W=589 um
S=S1 um
L=L1 um
W1=589 um
W2=611 um

W=611 um
L=9200 um

Z=50 Ohm

MSub

S-PARAMETERS

H=635 um
Er=9.8
Mur=1
Cond=4.1e7
T=12 um
TanD=0
Rough=0 um

Start=8 GHz
Stop=16 GHz
Step=0.01 GHz

L1=2800
S1=840
S2=1680
L2=1400

Fig. 1: Dimensions of the coupled lines filter at 12.3GHz

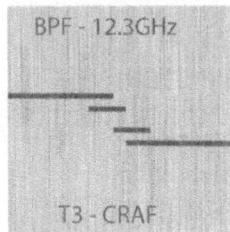

Fig. 2: Layout of the processed filter on alumina (scale1:1).

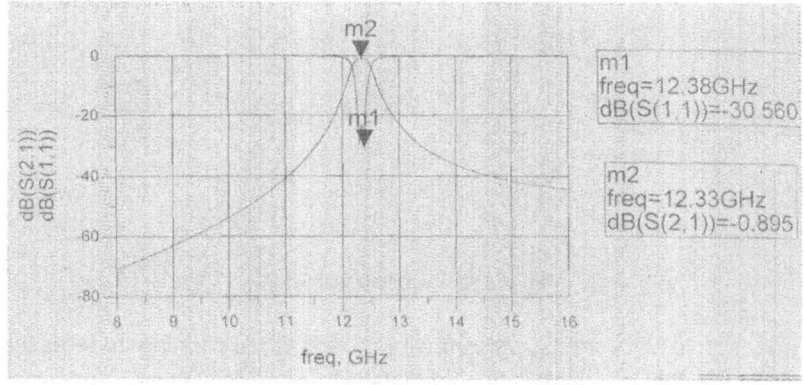

Fig. 3: Simulation of the filter using ADS

3.   REALISATION OF THE FILTER: THE CLEAN ROOM PROCESS

The interest was to choose a circuit simple enough to process it in a short time
(~ $2^{1/2}$ - 3 hours), allowing direct measurement after the clean room realization.
The main step was the photolithography. The different parts are further described.
First, cleaning of the alumina substate. Second, resist coating (with protection of
the back side ground plane). Third, the photolithography (UV insolation and development), including a short presentation of alignment techniques, to correctly
position the output patterns of the filter above the substrate. Then, Gold and Chro-
mium wet etching were possible, before resist stripping. Optical microscope controls could be done, to estimate the quality of the lines, and a final profilometry
measurement was added to know the exact value of the gold thickness, as well
as lateral dimensions of the lines. These values could be used in the back-simulation to evaluate the tolerances on the dimensions.
This very simple process was a good introduction to clean room work, the students being able to process their own circuit without particular training and with
100% success in the realization.

4.   MEASUREMENT OF THE FILTER

4.1   THE VECTOR NETWORK ANALYSER SET-UP

We used the HP8510B vector network analyser to measure the filter characteristics, with a microstrip test fixture, see Figure 4.

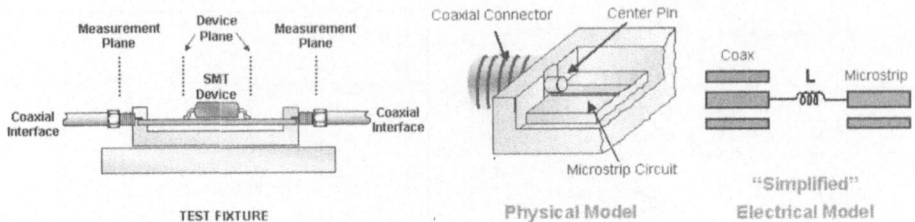

Fig. 4: The microstrip test fixture

An introduction to VNA measurement was proposed, explaining in detail the TRL calibration procedure and the associated specific errors (mismatching, frequency dependance,...). The synopsis of the VNA is proposed on figure 5.

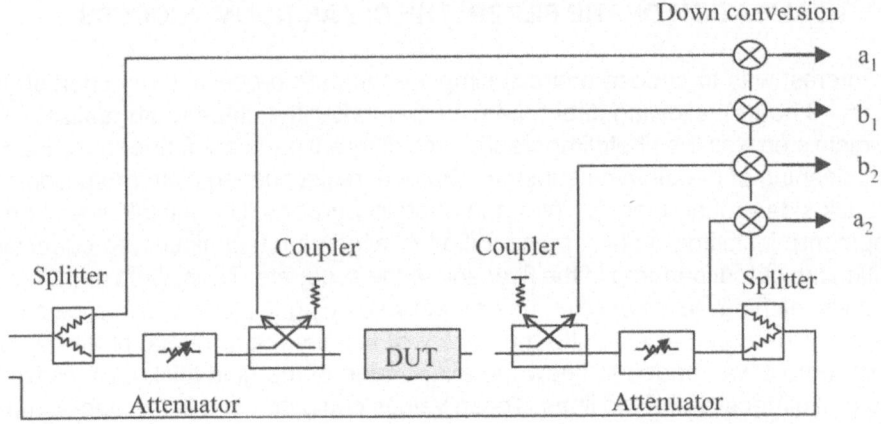

Fig. 5: Synopsis of the VNA

4.2   MEASUREMENT RESULTS

We obtained quite good results, with a frequency shift up to 12.45GHz – 12.6GHz, depending on the circuit, instead 12.3GHz, and more in-band losses than theoretically expected (~ 2-3dB instead of 0.8dB). The comparison was better with Momentum simulation results than ADS since it takes into account the metallic losses and the influence of the substrate.

5.   BACK-SIMULATION

5.1   INFLUENCE OF DIMENSIONS VARIATIONS

The idea was to try to explain some of the differences between measurements and the initial simulations with dimensions variations. Students could use the Tuning tool in ADS including the sensitivity on dimensions obtained after the profilometry step. A smaller value of width W could explain a higher frequency.

5.2   INFLUENCE OF THE SUBSTRATE

The same approach was possible considering the substrate, especially concerning the dielectric constant dispersion ($9.6 < \varepsilon_r < 9.9$), see Figure 6.

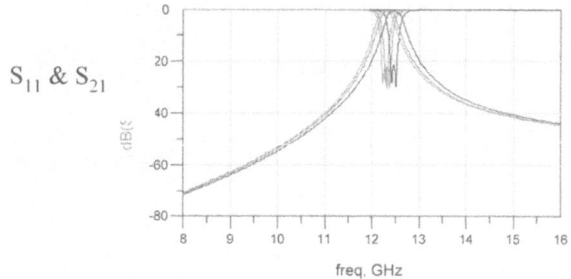

Fig. 6: Tuning on the dielectric constant

CONCLUSION

Such an experimentation was developed to illustrate a theoretical course on a specific RF technology, the coupled lines filters, generally difficult for the students. It was also correlated with an introduction to clean room process as well as RF measurement techniques, showing finally that the circuit obtained was close to the expected one after simulation. The whole study was splitted into a 3h tutorial using ADS simulations on different filters, then 3h in clean room, and at the end 2h around measurement and back-simulation.

REFERENCE

[1]   M. Villegas, "Radiocommunications numériques/2", DUNOD, 2002.

HOME APPLIANCE SYSTEMS AND DOMOTICS COURSE WITH MULTIMEDIA SUPPORT

GOMES L.[1], BORZA P.[2], COSTA A.[1],
lugo@uninova.pt, paul.borza@siemens.com, akc@uninova.pt

[1]*Universidade Nova de Lisboa / UNI-NOVA*	[2]*Transilvania University of Brasov*
NOVA	*Faculty of Electrical Engineering*
Faculdade de Ciências e Tecnologia	*Brasov*
Department of Electrical Engineering	*ROMANIA*
2829-516 Caparica - PORTUGAL	

1. Introduction

During the last decade, several attempts have been tried in order to integrate e-learning methodologies and multimedia capabilities for distance training in several domains. Hopefully, many of them succeeded. European Union's Leonardo da Vinci Programme is one very successfully Programme supporting the development of such methodologies and tools.

The present paper focus on some developments carried inside one Leonardo da Vinci project, named "Using information & communication technologies in development of virtual & remote laboratories for initial & continuous education oriented on efficient professional (re)insertion in electrical domain" (VIRTUAL-ELECTRO-LAB) [1], started 2002 till 2004.

Professionals already in the electrical domain that want to be acquainted with specific subjects covered by the project outcomes compose primary target group.

2. The VIRTUAL-ELECTRO-LAB project

The goal of the VIRTUAL-ELECTRO-LAB project is the development of a complex training system, including the correlation of courses, seminars, workshops and testing systems with virtual & remote experiment elements. This goal is supported through the usage of a virtual & remote laboratory, which can be remotely accessed through the Internet. For that, the following ICT products are under development: (1) The e-learning platform; (2) The virtual and remote laboratory; (3) The module courses (also called software tools).

The e-learning platform includes a web-site (http://demolab.ecampus.it), also supporting the access to the e-learning management system.

The virtual and remote laboratory is a key element of the project outcomes. It is supported by remotely controlled physical experiments (remote aspect), and also by remote access to specific simulators associated with specific experiments

A.M. Ionescu et al. (eds.), Microelectronics Education, 223–227.
© 2004 *Kluwer Academic Publishers.*

(virtual aspect). So, all partners (and registered users) can share the usage of existing advanced equipments and development environments.

The first module course is the preparatory module course, which main goal is to give information and specific training to the primary target group (persons who want to use the software tools developed by the project and do not have the necessary skills).

The nine module courses developed in the frame of the project cover the following topics: *Properties and characteristics of the electrotechnic materials, Simulation and computing of electrical circuits, The electrical transformer and the induction machine, Measurement of electronic devices and circuits, Simulation of electric drives, Home appliance systems and peripheral components (domotics), Web-oriented applications of databases used in electrical domain, Measurement & automated test system,* and *Simulation of the control systems used in electrical processes.*

Each module course is structured and prepared in order to be presented using different media and resources, ranging from a plain CD-ROM (to be used in a stand-alone fashion, like an electronic book), to an interactive tool accessed through the Internet with access to remote experiments and to the virtual laboratory.

3.   The Domotics course

The present paper focuses on the module course devoted to domotics. Initial structuring was based on a set of common thoughts, namely "What is the meaning for *domotics*", "What features for the *Dream House*", "Which are the integration goals", "What we can do with stand-alone and networked components at home", "How to put together residence solutions and institutional building solutions", and "Advantages of current technologies when applied in house".

Resulting contents is the following (using the "credit" concept to roughly identify the relative length of the chapter; total credits: 20):

Chapter	Title	Credits
1	Introduction to domotics	3
2	Comfort management systems	2
3	White goods and Entertainment integration	2
4	Load management in house	2
5	Access, Security and Safety Systems	3
6	Communications in house	2
7	Protocols & buses used in house	2
8	Interfaces for home appliance systems	1
9	Programs for applications	1
10	The Intelligent Building concept	2

Without going into details (due to space limitations), it is clear that we tried to cover all major appliances presented in houses, and also making the bridge with institutional building systems, through a presentation associated with the "intelligent building" concept.

Obviously, one aspect of the module course is to present a hypertext document regarding each of the referred subjects. Complementary emphasis is put on the interaction with remote experiments, in several ways, as described in the following section.

4.   Some experiments and their exploitation

With the set of experiments under preparation, we intend to reach different levels of exploitation:
- First level is based on the remote control and monitoring of already available systems.
- Second level foresees (limited) configuration, reconfiguration or/and reprogramming of specific sub-systems.

Clearly, first level is associated with "regular" users, supporting remote operation and "play for fun", while second level supports intermediary and advanced topics where the users can change configuration (at some extend).

We take the access control system as one example. At the first level (introductory level), one can only remotely monitor the activity of the different access control controllers. Although, at the second level, one can change configuration of specific access control sub-systems in several ways; for instance:
- We can change the users' database characteristics, namely users and associated privileges (intermediate level),
- We can change parts or the whole controlling program, as far as the system supports dynamic reconfigurability (advanced level). This feature will allow the user to download specific control strategies to the controller. As a specific example, one can think about a different algorithm to be used for biometrics identifycation, for instance, through hand-geometry or fingerprint recognition (just to mention a few). This dynamic reconfigurability is supported by an adequate RTOS (real-time operating system) installed in the specific sub-system.

As a matter of fact, the already referred introductory, intermediate and advanced levels are concepts already applied to the theoretical part of the course, namely to the associated hypertext, to allow some guidance to the users according to their expertise, motivations and goals.

As a second example, we can consider the heating and air-conditioning system: at the introductory level, the user can only monitor house variables for every room (temperature and humidity); at the intermediate level, the user can change the set points and operative modes (week-days, week-ends, holidays, and so); at the advanced level, the user can change everything, even the control algorithm (from a

PID controller to a fuzzy controller, for instance). Afterwards, the user can access to the variable evolution registered in the database and conclude about the effectiveness of their control strategies. This is accomplished through the capabilities offered by the remote and virtual laboratory, close to the one proposed in [2] .

4.1 NETWORKED CONTROLLERS

As an example of a specific set-up supporting remote experiments, we present a system composed by networked controllers, nick-name TinyDomots (*TinyDomoticRobot*), that can be remotely monitored and configured (using the facilities of the remote and virtual laboratory), as shown in Figure 1 - Topology of the netwrok for home appliances control 1. Each TinyDomot can have a simplified user interface (keyboard and display), which will be reproduced remotely, several inputs from sensors (temperature and light sensors, presence detectors, and so), and outputs to actuators (lamps, fan, several equipments), and can be equipped with one network interface. Several communication supports are possible, from high-speed LAN (Ethernet), to dedicated low-speed LAN (RS-485), wireless and through the power-line. Specific adaptors making the bridge between the different sub-networks are available (bridging wireless and power-line, and so on). All controllers and adaptors were developed using low-cost micro-controllers, with free-of-charge or low-cost development environments, in order to allow their usage by a wide number of users.

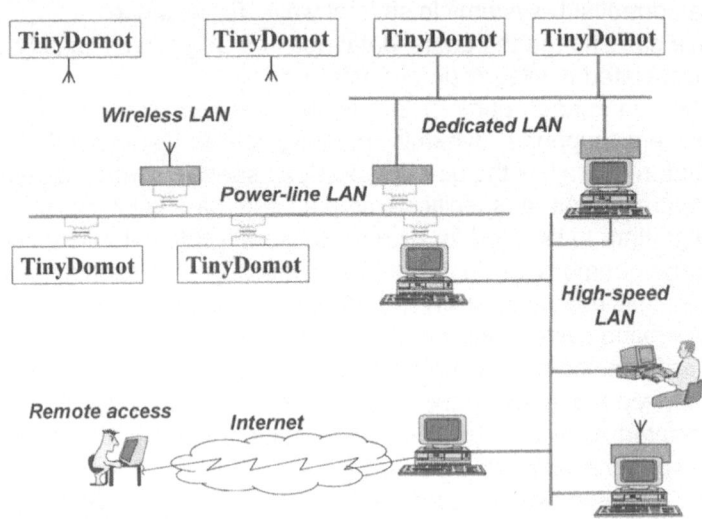

Fig. 1: Topology of the network for home appliances control

Conclusions

The referred course will be used for initial & continuous education & training on domotics, by people with or without previous experience on the subject. It will also be used as an experimental test-bed for specific embedded systems development training environment, namely for the ones associated with the most common sub-systems presented at houses.

References

[1] VIRTUAL-ELECTRO-LAB – Proposal for a Pilot Project - RO/01/B/F/ PP141024

[2] E. Kayafas, F. Sandu, I. Patiniotakis, P.N. Borza, "Approaches to Programming for Tele-Measurement", Proceedings of the XVII World Congress of IMEKO – International Measurement Confederation – Lisbon, 2001; selected to be published in a special volume by Elsevier

INDUSTRY RELATIONS

ORAL PRESENTATIONS

A MODEL FOR UNIVERSITY-INDUSTRY COLLABORATION

The Center for Analog and Mixed Signal IC Design at WPI
MC NEILL J.-A.
Worcester Polytechnic Institute, 100 Institute Rd. Worcester, MA, U.S.A
mcneill@ece.wpi.edu

ABSTRACT
This paper describes the New England Center for Analog and Mixed Signal Integrated Circuit Design (NECAMSID). This center, in the department of Electrical and Computer Engineering at Worcester Polytechnic Institute (WPI), enables collaborative interaction between industry sponsors and WPI faculty. The flexibility of WPI's project-based educational program allows a variety of undergraduate and graduate educational opportunities, including industry sponsored projects at both on- and off-campus sites.

1.   INTRODUCTION

Despite recent volatility in the field of electrical and computer engineering, there is a persistent need for analog and mixed signal design engineers [1]. The needs of industry require designers to understand a broad range of disciplines, from the solid-state physics of submicron devices, through circuit-level techniques, to high-level system design and analysis. In addition to theoretical knowledge, the designer must also be constantly aware of manufacturability and testability issues. The increased pressure on educators and students to squeeze more information into the curriculum motivates the exploration of enhanced of university-industry interaction.

This paper describes the New England Center for Analog and Mixed Signal Integrated Circuit Design (NECAMSID). This center, formed in response to the needs of industry, student, and faculty constituencies, is one model for collaborative university-industry interaction. This paper is organized as follows: after an introduction to Worcester Polytechnic Institute (WPI) in section 2, the NECAMSID model is described in section 3. Examples of undergraduate projects and graduate thesis work performed are given.

2.   THE UNIVERSITY

2.1   WPI'S PROJECTS PROGRAM

Founded in 1869, Worcester Polytechnic Institute is the third oldest private school of engineering and science in the United States. WPI has about 220 faculty, 2600 undergraduates, and 600 graduate students, and offers programs in engineering,

A.M. Ionescu et al. (eds.), Microelectronics Education, 233–237.
© 2004 *Kluwer Academic Publishers.*

the sciences, and management. In 1970, WPI adopted "The WPI Plan", an innovative undergraduate curriculum which focuses on outcome-oriented, project-based education. In addition to other degree requirements, all WPI undergraduates must complete two significant projects in order to graduate: the Interactive Qualifying Project (IQP) and the Major Qualifying Project (MQP). Both of these projects are done in small groups (typically three students) and in close collaboration with one or more faculty advisors.

The IQP, typically done in the junior year, is an interdisciplinary project in which student teams address a problem relating technology and society. The MQP, typically done in the senior year, is a disciplinary project which constitutes a "capstone design experience" or "senior thesis" in the student's major area. Both of these project experiences emphasize the development of skills related to teamwork, open-ended problem solving, and oral and written communication. Almost half of all WPI undergraduates complete one of these projects at an off-campus site. WPI has IQP project programs in such places as Washington, San Francisco, Puerto Rico, Costa Rica, London, The Netherlands, Germany, Venice, and Bangkok. Senior projects are conducted at project sites such as Limerick, Ireland, NASA's Goddard Spaceflight Center, and the Silicon Valley Project Center. Students spend an academic term (two months) at one of these sites completing their project under the on-site supervision of a WPI faculty member.

2.2   ECE DEPARTMENT CURRICULUM

Students must acquire a broad range of skills to succeed in this project-oriented program. These are the same skills that our graduates will need in an increasingly competitive global marketplace: high personal motivation, flexibility and breadth, critical thinking, creativity and experimentation, ability to work in teams, ability to communicate effectively (both verbally and in writing), and ability to learn independently

The project-oriented philosophy is present throughout the curriculum, beginning with introductory courses that give an overview of the entire ECE field, and continuing to advanced courses that deliver targeted knowledge required to perform advanced project work.

The entry point of the ECE curriculum is EE2011, "A Project-Oriented Introduction to Electrical and Computer Engineering." [2] This course is usually taken by students at the end of the first year, and as the title indicates, the emphasis is on a hands-on, project-oriented introduction to the field of electrical and computer engineering. Once students encounter this overview of the engineering profession, they are better motivated to learn the basic principles and inner workings of the systems they have seen.

The project orientation continues with a recently introduced design course, EE2799, in which students are encouraged to think at a system level: what kind of design problems are addressed by electrical and computer engineers, and how are these problems solved? Typically taken in the sophomore year, this course provides an early introduction to design principles [3, 4]

For students who wish to perform their MQP in the area of mixed signal integrated circuit design, the curriculum has recently been restructured so that the advanced analog integrated circuit design course, EE4902, is available at the end of the junior year. In this way, students entering the senior year will be able to complete a mixed signal integrated circuit design as their MQP project. Additionally, the delivery of this course has been modified to a studio format, which has had a positive effect on student learning [5]

2.3   RESEARCH FACILITY: MIXED SIGNAL DESIGN RESEARCH LAB

A major part of the ECE department's commitment to project-based education is the Mixed Signal Design Research Laboratory. The laboratory facility comprises a complete integrated circuit design and test environment, and is an integral part of the project experience in analog and mixed signal integrated circuit design. Funding from industry partners and the National Science Foundation was used to equip the lab with workstations, CAD software, and test instrumentation. Integrated circuit fabrication is available through MOSIS and the industry partners. The result is a facility that allows project students to experience the complete circuit design and simulation process: schematic capture, simulation, layout, parasitic extraction, layout-vs.-schematic verification, fabrication, test, and evaluation. Students participating in on-campus projects have full access to this lab. The goal is an education that produces a design engineer who is knowledgeable in all stages of the design process including manufacturing, testing, and meeting the end customer's application requirements.

3.   THE COLLABORATIVE CENTER

3.1   PURPOSE

The purpose of the NECAMSID Center is to take advantage of the common themes that link mixed signal design projects, and to create an environment where all industry sponsors can take maximum advantage of their common interest in mixed signal integrated circuit design. The goal is an environment in which information, and especially contact with students, can flow freely among all members.

Note that this center is intended to complement targeted research, not replace it. An open, collaborative model is not appropriate for investigation of research topics involving proprietary concepts. The Center faculty and laboratory facilities continue to be available for separately funded, company-specific research.

The Center is supported by membership fees which are contributed by microelectronics companies with a presence in New England. Research and project topics are proposed by both Center faculty and by the member companies. An industry Advisory Board, composed of representatives from the member companies, makes the final decision regarding which projects will be supported.

3.2   BENEFITS

There are numerous benefits for the member companies:
Access to graduating seniors, M.S. students: Each year there are about 15 students working in the lab, on a combination of research and project work in mixed signal IC design.
Increasing pool of students with mixed signal IC design experience: Experience has shown that an active research center creates an atmosphere of excitement and student interest in the center's field. Even for students who are not working directly in the lab, there will be an increase in interest for mixed signal IC design.

More awareness of sponsor's company among all students in ECE: Similarly, an active research center creates an atmosphere of interest about the companies who are sponsoring research and project work. This helps to create a positive "word-of-mouth" among students.
Influencing direction of research: For member companies with an expertise in specific subdisciplines of mixed signal design, membership in the Center offers an opportunity to influence the choice of research projects, and thus the direction of future research.
Awareness of and access to new technologies: Participation in the Center's research activites is a quick, low-cost way for member companies to get "up to speed" in other new areas of technology development
Influencing curriculum development: New graduate courses are constantly under development to serve the needs of ongoing research in the department.
Networking: Twice-yearly meetings offer members a chance to get together with colleagues from other companies.

3.3   EXAMPLE PROJECTS

Since its inception in 1997, more than 30 undergraduate projects and 10 MS theses have been completed in the Center. More information on specific projects is available at http://www.ece.wpi.edu/analog and in publications resulting from previous project and thesis work [6-8].

ACKNOWLEDGMENT
The author acknowledges the support of the National Science Foundation, which provided funding under grants MIP-9701408 and CDA-961733, and the valuable advice of Prof. Terri Fiez of Oregon State Univerisy.

REFERENCES

[1]   J. McNeill and R. Vaz, " Proceedings of the 1997 IEEE Computer Society International Conference on Microelectronic Systems Education (MSE'97), Arlington, VA, July, 1997.

[2]   J. McNeill and R. Vaz, "High Expectations: A Passport to Success," Ninth International Conference on the First Year Experience, July, 1996.

[3]   R. Vaz, S. J. Bitar, T. Prestero, N. Cantor, "Student Design for the Developing World," Proceedings of the 2004 ASEE Annual Conference & Exposition, Session 2260

[4]   R. Vaz, "A Sophomore-Level ECE Product Design Experience,"Proceedings of the 2004 ASEE Annual Conference & Exposition , Session 1725.

[5]   J. McNeill and K. Keenaghan, "Transitioning an Engineering Course to Studio Format," Proceedings of the 2002 IEEE Frontiers in Education Conference (FIE '02), Boston, MA, October, 2002.

[6]   J. McNeill, M. Lawler, G. Levesque, J. Ruiter, J. Noon A 50A, 1-us-rise-time, programmable electronic load instrument for measurement of microprocessor power supply transient performance , Proceedings of the 2000 IEEE IMTC2000, Baltimore, MD, May, 2000.

[7]   J. McNeill, "Boost Converter Provides Temperature-Controlled Operation of 12V Fan from +5V Supply", Design Idea, EDN Magazine, pp. 98-100, December 18, 1997.

[8]   J. McNeill, J. Kulesza, L. Menezes, N. Tjoa, "Improved Efficiency Fan Speed Control," Ideas for Design, Electronic Design Magazine, pp. 98-100, October 22, 1998.

AN EDUCATION-ORIENTED INDUSTRY AND UNIVERSITY COLLABORATION: A CASE STUDY OF LG-HGU MODEL IN KOREA

YI K*., JUNG K.-H.*, SUNG K.-Y.*, LEE D.-H.**
*School of Computer Science and Electronic Engineering, Handong Global University, Pohang, Gyungbuk, 791-708, Korea
**Digital Display & Media Company, LG Electronics, 642 Jinpyoung-dong, Gumi, Gyungbuk, Korea

ABSTRACT

We present a new collaboration model between industry and university. The main strategy is to develop demand driven courses to bring the required field knowledge into the conventional classes. It's a kind of win-win strategy for both academia and industry to have short-term benefits as well as long term benefits. We have applied the proposed model to collaboration between the Digital TV Company of the LG Electronics and Handong Global University in Korea.

1. INTRODUCTION

From the view point of Industry who is the main consumer of the product of university engineering education – the people trained for engineering work, one of the most important aims of the engineering education is to prepare people as problem solvers who are equipped with practical methods and understanding for the real problems in the field. But, most of the engineering education does not meet the expected level of industry in terms of problem-understanding and -solving capabilities. Thus, most of the companies re-train the new employees who graduate the engineering school from the basic level to the focused special topics each company is interested in as well as the industry-specific spirit education. It reflects the industry does not trust university education because of the different direction and focus from those of fields. It cause a vicious circles : industry does not trust the university education and the class quality is lowered. In order to break the vicious circles, courses designed and operated by both industry and university is required as proposed.

To minimize the gap between the field knowledge and education contents, the collaboration between industry and university has been emphasized for a long time. Although the importance of the collaboration is well understood by the most of industry and school administrators, conventional collaboration models have failed to obtain the desirable results because in most cases the collaboration was focused just on small number of research projects and the company just expects a few people who involved in the research. It is a very limited collaboration model and cannot contribute to the improvement of undergraduate course work for practical industry knowledge.

A.M. Ionescu et al. (eds.), Microelectronics Education, 239–242.

Here, we propose a new cooperative model between industry and university to bring the classical courses the field knowledge and increase the effectiveness of university education. Our model is education-oriented and demand-driven education rather than research project-oriented. With this approach all related parties - students, faculty, industry, and academia get benefits.

2.   A NEW STYLE OF COLLABORATION BETWEEN ACADEMIA AND IN-DUSTRY : A WIN-WIN MODEL

Figure 1 shows the collaboration model centred on the costs to pay and benefits expected.

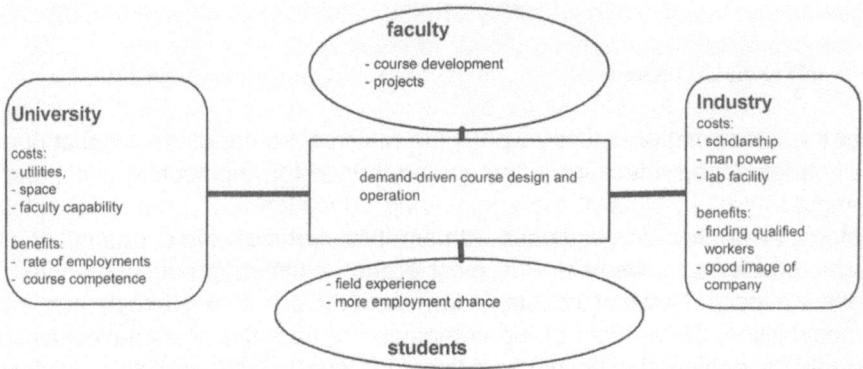

Fig. 1: LG-HGU Cooperative Model : cost and benefits for each part

University and students have benefits from the collaboration in forms of higher quality recruitment, increased rate of employments, enhanced course development with most updated field knowledge.
The main features of the new model are as follows:
- Education-oriented model rather than research-oriented;
- Covers undergraduate and master courses;
- Co-developing courses and team-teaching by both industry and academia;
- Focused on the quality of the university education to meet the demand of industry; and
- Establishing beneficial circle the effectiveness of engineering education and supplying the qualified employee to industry

The proposed model was applied for collaboration between LG Electronics Digital TV software development group and Handong Global University. The aim of the collaboration is to train and to prepare the students with digital TV system field knowledge.

3.   DEVELOPEMENT OF COOPERATIVE COURSES

The heart of the education-oriented collaboration model is the new course developed and operated by both university and company. Handong Global University (HGU) and LG Electronics (LGE) implemented a project-based senior course as a starting point of the education-oriented cooperative model. Thus, it was necessary to prepare a new laboratory environment in order to operate this course. The classroom with computers, and some instrumental devices are provided by HGU, and the PDP TV sets, audio system, beam projector, and software tool set including compliers are donated by LG Electronics.

Meanwhile, HGU appointed the several research engineers of LG who would give lectures for this course as the adjunct professors. Not only the LG engineers but also HGU faculty members operate the class in order to supervise the experiments and lab projects and to find and solve any obstacle against the improvement of this course

There were several meetings and a workshop to discuss the scope and purpose of this collaborative course. The main topic includes digital TV system and its embedded software, especially graphical user interface and communication protocol to control the TV set. The digested course syllabus is given in Table 1.

Table 1. The outline of the course syllabus

Topics	number of Weeks	Project	Remarks
Introduction to the TV system	2		
TV Graphic User Interface	4	Graphical User Interface Using Paradigm Complier	
RC232C Protocol	2	Remote control by RS-232C	
Inter IC Communication	2		
Field Observation	1		LG DDM Company
DTV System	3	DTV Parsing	
Project Presentation	1		Project Contest

Since it is considered to be very helpful for students to experience the industry filed where they make use of what they learn in the course, the opportunity of field experience to the LG DDM Company in Gumi is scheduled during the course-work. The course is essentially project-based class and three term projects are planned as shown in the Table 1. And there would be a contest for the project out-puts at the end of the coursework to promote the motivation of the students.

In order to settle down and upgrade this kind of new course, it is indispensable to evaluate the course. Because this course is currently going on, there is only inter-mediate feedback from student. However, the assessment will be performed in three aspects, that is, from the viewpoint of university, industry and the student who is an actual consumer of education. After the course is finished, there would be a meeting between HGU and LG to evaluate the achievements of this course and to discuss the expansion plan of the width and depth our collaboration model. One of the positive finding is that many students in this course succeed in the campus recruit program of LG and will start their first carrier as a TV engineer. Therefore, we expect it will be possible to evaluate their adaptability and to show the assessment result at the conference in June, 2004.

CONCLUSIONS AND FUTURE PLAN

In this paper, an education oriented collaboration model between industry and university is presented. Our main concern is to develop a win-win model for both industry and university as well as to bring up a competent student who can do his or her own job without any re-education at industry. We anticipate the introduced cooperative course that HGU-LG co-developed would be a cornerstone to keep the fruitful partnership.

HGU and LG want to enlarge and deepen the relationship so that this will be a promising cooperative model between engineering academia and industry. Our future plans are as follows:

- To develop intensive internship course for senior student;
- To open another cooperative course (TV display or embedded SW);
- To increase the LG scholarship program;
- To find education oriented research topics; and
- To formalize the workshop between the faculty members of HGU and the research engineer of LGE.

REFERENCES

[1]   B. Lee, T, Noh, C. Kim, "The Plan for Future-Oriented Assessment of Engineering Education", the Journal of Korean Society for Engineering Education & Technology Transfer, VOL. 04, NO. 01, pp. 3- 17 June 2001.

MULTIMEDIA AND INNOVATIVE TEACHING METHODS

ORAL PRESENTATIONS

MULTIMEDIA TOOL FOR UNDERSTANDING AND EXPLAINING MICROSYSTEMS

FISCHER P., ECABERT M.
FSRM, Ruelle DuPeyrou 4, CH - 2001 Neuchâtel, Switzerland
+41 32 720 09 00, fischer@fsrm.ch
BERGER M.
mib génie logiciel, Neuchâtel, Switzerland
MOUNIER E.
YOLE Développement, Lyon, France
BOTTHOF A.
VDI/VDE-IT GmbH, Teltow, Germany

1.   INTRODUCTION

This paper presents the work accomplished in the IST project CD-MST, the final deliverable of which is a new interactive CD-ROM presenting the world of micro-systems to different target groups – from students to venture capitalists.

The partners of the project are FSRM - Swiss Foundation for Research in Micro-technology, in Neuchâtel, Switzerland, VDI/VDE-IT, in Teltow, Germany, YOLE Développement, in Lyon, France and mib - génie logiciel, in Neuchâtel, Switzer-land.

The data on the CD-Rom is of interest to education (engineering students and ac-ademics) on account of its pedagogical approach, industry, finance and the me-dia. It highlights a large number of industrial achievements, the associated technology and current applications, as well as detailed economic analysis.

The work is based on the successful CD-ROM edited in the year 2000 by FSRM and presented at EWME 2000, in Aix-en-Provence.

2.   ORGANIZATION OF THE CD-ROM

As it appears on the left side of the home page (figure 1), the CD-ROM has three main entries:
*   **Markets**, with an extended description of 8 major market areas of microsystems.
*   **Technologies**, with in depth descriptions of 27 major fabrication processes used in microsystems technologies.
*   **Products**, with an extended description of 10 industrially manufactured examples of microsystems.

The three parts are highly interconnected through many hyperlinks.

A.M. Ionescu et al. (eds.), Microelectronics Education, 247–251.
© 2004 *Kluwer Academic Publishers.*

The CD contains over 230 illustrated screens, of which 70 are animated and include audio tracks with spoken comments. A very fast glossary is included with more than 150 entries and many players are mentioned with internet links.

The complete data, including the audio tracks, are available in English, German and French and the language can be switched at any time and any place in the CD.

A highly user-friendly navigation concept was developed to provide an optimal use of the large amount of information included in the CD. At the end of a visit, the user can have an overview of the already seen pages and the remaining pages to visit.

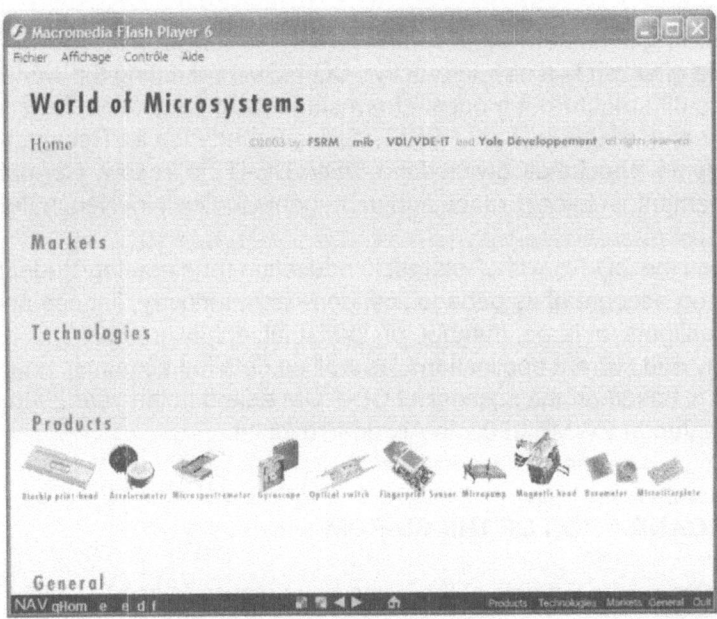

Fig. 1: Home page of the CD-ROM "World of Microsystems", showing the main entries: Markets, Technologies and Products

3. The content

The content of the CD is well suited for self study, for explaining and for promoting microsystems.

3.1   MARKETS

A selection of 8 major markets of microsystems was made. For each of them, the user can find following information:
- why microsystems are used in the market,
- the main characteristics of this particular market, such as volumes, margins, regulations and constraints,
- market data today and in 2005

and
- a list of major actors, with internet links.

A total of 32 illustrated screens are devoted to the markets and provide a good understanding of what the potential market penetration of microsystems can be. This part is particularly important for the promotion of microsystems.

3.2   PRODUCTS

A selection of 10 microsystem products was made, according to following criteria:
- a good representation of all major markets,
- a high degree of maturity: the products are industrially manufactured,
- a good representation of the major fabrication processes
- and finally
- the willingness of the manufacturing company to provide sufficient information on their processes and on economic aspects

The final list of selected products is shown is table 1.

Product	Market	Technology
Biochip printhead	Biotechnology	Plastic replication
Accelerometer	Defence and aerospace	Bulk micromachining
Microspectrometer	Instrumentation	LIGA + plastic replication
Gyroscope	Automotive	Surface micromachining
Optical switch	Telecommunication	Surface Micromachining
Fingerprint sensor	Security	CMOS
Micropump	Medical devices	Bulk micromachining
Magnetic head	IT peripherals	Thin films
Barometer	Consumer	Bulk micromachining
Microtiterplate	Biotechnology	Plastic replication

Tab. 1. List of selected products with corresponding markets and fabrication technology

For each of the products, the user can find following information:
- a short description,

- an extensive description of how it works, with animation and spoken comments,
- an extensive description of the fabrication process with animation, spoken comments and hyperlinks to detailed descriptions of each process step
- a overview of the major applications of the product, often with animation and spoken comments
- and finally
- an extensive economic analysis, including the production volume, value chain, roadmap, competing technologies and competitors (with internet links.

A total of 76 illustrated screens and 56 minutes of animation are devoted to the products and provide a good overview of what microsystem products are today and where they are used.

3.3   TECHNOLOGIES

All major fabrication processes are presented in the CD. They are accessible either directly or with hyperlinks through the fabrication descriptions of the products. In total, 27 fabrication processes are presented. They include the major silicon processes, but also LIGA, plastic replication, electroplating, laser micromachining and EDM.

For each process, the user can find following information:
- a short description,
- the purpose of the process, related to microsystem technology,
- an extensive description of how it works, with animation and spoken comments,
- a description of the necessary equipment, with a rough figure of the costs and a mention of the major manufacturers, with internet link.

A total of 105 illustrated screens and more than one hour of animation are devoted to the fabrication processes and provide an in depth description of microsystem technologies.

4.   USE OF THE CD-ROM

The main use of the CD-ROM is for education purpose. Students in microsystem technology can get a realistic overview and a complete summary of microsystems technologies and applications. It provides highly valuable information complementary to that of lectures and books.

Many professors also use the CD-ROM to illustrate and complete their lectures. A very useful tool is included to edit dedicated slideshows, in which personal pages can be inserted, with comments and titles.

Finally, many promotion, interface and technology transfer organisations use the CD-ROM for promotion purposes, also making use of the slideshow function.
One month after its launch, over 400 copies of the CD-ROM "World of Microsystems" have been sold, worldwide

REFERENCES

[1] Philippe Fischer and Annette Locher, "Continuing Education in Microsystems Technology: FSRM's 10 Years of Experience", mstnews No 5/03, pp8-10.

[2] Philippe Fischer and Annette Locher, "Continuous Education in Micro- and Nanotechnology", Proceedings of COMS2003.

[3] Philippe Fischer, Johachim Rupp, Alfons Botthof, Eric Mounier, Sandrine Leroy, Michel Berger, "Business Model Analysis of Commercialized Microsystems Products", Proceedings of COMS2003.

OVERVIEW OF E-LEARNING ENVIRONMENT FOR WEB-BASED STUDY OF TESTING AND DIAGNOSTICS OF DIGITAL SYSTEMS

Extended Abstract
JUTMAN A.[1], UBAR R.[1], WUTTKE H.-D.[2]
[1]*Tallinn Technical University, ATI, Raja 15, 12618 Tallinn, Estonia*
[2]*Technical University of Ilmenau, Helmholtzplatz 1, 98693 Ilmenau, Germany*

1.   Introduction

In this paper, we present an overview of latest developments taking place at Tallinn Technical University (TTU) in the area of e-learning supported by European project REASON (REsearch And Training Action for System On Chip DesigN) [8].

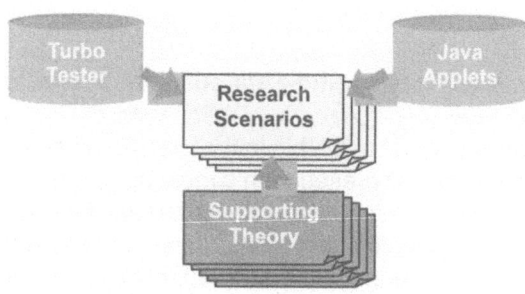

Fig. 1: Overview of the e-learning environment

The environment consists of several interrelated modules shown in Figure 1. The PC-based tools installed locally and Java applets invoked remotely via Internet supplement each other and form the engine of the whole concept. The PC-based tool set is called Turbo Tester (TT) [1,7] and consists of the following main tools: test generators based on various algorithms, logic and fault simulators, a test optimizer, a module for hazard analysis [6], a simulator and test generator for defects [2], built-in self-test simulators [4], design verification and design error diagnosis tools [5]. This range of compatible diagnostic tools forms, via their interaction and complementary operation, a homogeneous research environment, which provides good possibilities for laboratory training and experimental research.

The general idea behind the Java applets is a bit different. They mainly aimed at supporting the concept of game-like style of learning via easy action and reaction, learning by doing, and concentration on most important topics in the simplest possible way. There are three applets available by now [9]. The fourth applet, which

A.M. Ionescu et al. (eds.), Microelectronics Education, 253–258.

represents a simple schematic and decision diagram (DD) editor, is currently under development.

The central point of the presented concept is research scenarios – a set of problems and experiments that represent virtual laboratory works, where students learn diagnostic software and get acquainted with common concepts and problems of testing and diagnostics. There are scenarios for beginners and for advanced study. The user-friendly graphical environment created by the applets is the best option for beginners. More advanced study is possible with scenarios, which make use of TT tools. The full text of the scenarios is available in the Web [10].

The rest of the paper is concentrated on the description of the applets and related scenarios leaving the PC-based part of the e-learning environment behind the scope of the article.

2.   Overview of Java Applets and Related Work Scenarios

There are three applets already available in the Web [9] and one more is currently under development. Figure 2 represents an overall structure of relations between the applets, Turbo Tester, and the research scenarios. As it is seen from the figure, the scenarios form two distinct groups: Java applet-based (and therefore available on the Web) ones and TT-based (advanced) ones. Several advanced scenarios also make use of some functionality of the applets. Another group of scenarios fully exploit the functionality of the corresponding applet and therefore each such applet-scenario pair represents a self-contained system aimed at teaching target area of knowledge and engineering. We tried to design our applets in a uniform way; so that the user once got acquainted with the overall style does not have to spend his time learning the new style once again from the beginning. In the following we describe the functionality of the applets in more detail.

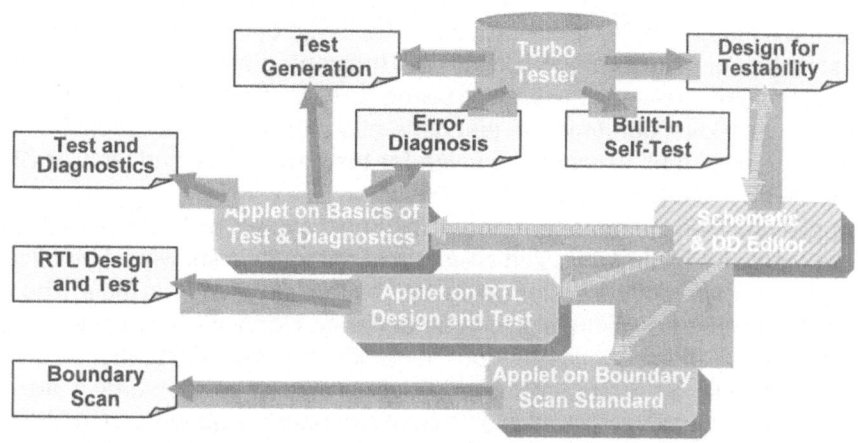

Fig. 2: Relationship between teh applets, TT, and research scenrios

Java applet on basics of test & diagnostics. Main principles of logic-level test generation and fault diagnosis are the fundamentals of the modern VLSI CAD and diagnostic software. Good understanding of these basics gives students vital background and skills in their future profession. The applet provides with possibility of manual and pseudo-random test pattern generation (TPG), fault simulation, combinational and sequential fault diagnosis, and investigation of BIST techniques. All the designs used in the applet are combinational circuits of a rather small size represented and visualized on the logic level.

In the test generation mode one chooses a target fault and step by step activates needed paths in the circuit in order to detect the fault. At the same time, the lines (wires) change their color helping the student in selection of proper values. If the specific fault detection is not of a primary interest, the user can specify the input values only. In this way one can set up as many vectors as he wants. The pseudo-random TPG is also implemented. It provides two modes: BILBO and CSTP [3].

All the vectors can be then simulated in the fault simulation mode. The results of simulation are represented in the form of fault table. By selecting a vector in the table, all the faults detected by this vector will be highlighted on the design schematic. In the combinational fault diagnosis mode, a subset of vectors is selected and applied to the erroneous circuit (imitating test experiments). The applet shows the results of fault diagnosis by highlighting the faulty area. If the results are not satisfactory, additional vectors could be generated.

Another supported method of fault diagnosis, the sequential diagnosis is based on the guided probing strategy. In this mode, students should sample the values by clicking signal lines and comparing observed values and correct ones. The goal is to find the precise fault location using as few samples as possible.

Java applet on RT-level design and test. In this applet, we combine and illustrate many different problems related to RT-level control intensive digital design, test, and design for test. Therefore, the applet gives a unique possibility to teach these topics in a consecutive iterative approach. The range of considered problems includes: design of data path and control path (microprogram) on RT level; investigation of tradeoffs between the system speed & HW cost; RT-level simulation and validation; gate-level deterministic test generation for RT-level functional blocks, functional testing, fault simulation, design for testability, logic and circular BIST, functional BIST, etc.

The applet provides the representation of the target system on RT level as divided into *datapath* and *control unit*. The structure of datapath is shown schematically, while the *control part* of the system is defined by the *Microprogram table*. The RT-level fault and fault-free simulation can be performed for a single set of input data as well as for all the sequence of input operands at once. During the simulation the active line of the microprogram is highlighted. The simulation data is also reflected in a graphical way. There is a manual test patterns generation mode for a selected microoperation separately. The gate-level representation of this micro-operation is shown at that time. The test access to the selected unit is provided by a special *Test Microporogram*. There are various possibilities to experiment with BIST provided in the *BIST module*. The user can select locations of pseudo-random TPGs and signature analyzers within the data path as well as their con-figurations.

The applet has a flexible design. The RT-level system model is described in a form of text-files. Hence, any custom design can be described and loaded just as simple as the original ones. The applet has a built-in extendable collection of examples implementing different algorithms. For connecting the applet to other applications as well as for allowing users to save the results of their work, the applet has a data import/export capability. It also has a built-in multilingual support.

Java applet on Boundary Scan (BS) standard IEEE 1149.1. There are two different supported modes of BS simulation. The first one, the *TAP Controller Mode*, provides a very detailed illustration of operation of BS registers and the TAP con-troller. This mode is intended for the beginners and for teachers. It helps to understand all the needed basics. Another mode, the *Command Mode*, can be used for faster simulation of BS commands like EXTEST, SAMPLE/PRELOAD, etc. with different predefined input data. This mode is useful for *fault diagnosis*. For that, the applet supports possibility of random or specific fault insertion. The operation of the faulty device can be then simulated and the fault can be diagnosed. Our applet is provided with lots of built-in examples. Furthermore, it allows users to generate their own examples by creating fully custom chips or boards. In the chip editing mode, the applet reads the description of BS structures using BSDL (Boundary Scan Description Language) format which is a part of the BS standard. Such BSDL descriptions are usually free of charge and widely available via Inter-net, which makes the work with the applet easier and more exciting, since the student can visualize behaviour of many different chips available in the market.

Conclusions

In this paper we have described Java applets used in the e-learning environment developed at Tallinn Technical University. This work is still in progress. All the components of the environment are available via the Web.

The core of the system is a set of research scenarios or virtual laboratory works based on two types of software: *a)* diagnostic package Turbo Tester and *b)* set of Java applets. While the TT should be installed on a local computer, the applets are invoked remotely via Internet. The Turbo Tester provides a homogeneous research environment, which allows for interesting experimental research to be conducted. The Java applets, in their turn, provide an attractive game-like milieu, which is important especially for beginners.

Due to the facts above, the conception, overview of which is presented here, is suitable for a broad audience of learners who are interested in studying different concepts of testing and diagnostics of integrated digital circuits.

Acknowledgements

This work was supported partly by the EU Framework V project REASON, by the Thuringian Ministry of Science, Research and Art (Germany), and by the Estonian Science Foundation Grant No 5649.

References

[1] M.Aarna, E.Ivask, A.Jutman, E.Orasson, J.Raik, R.Ubar, V.Vislogubov, H.-D.Wuttke. "Turbo Tester - Diagnostic Package for Research and Training," *in Scientific-Technical Journal "Radioelectronics & Informatics"*. KNURE. Vol. 3(24), 2003, pp.69-73.

[2] M. Blyzniuk, FT. Cibakova, E. Gramatova, W. Kuzmicz, M. Lobur, W. Pleskacz, J. Raik, R. Ubar. "Hierarchical Defect-Oriented Fault Simulation for Digital Circuits," *IEEE European Test Workshop*, Cascais, Portugal, Mai 23-26, 2000, pp.151-156.

[3] M.L. Bushnell, V.D. Agrawal, *Essentials of Electronic Testing for Digital Memory and Mixed-Signal Circuits*, Kluwer Academic Publishers, Dordrecht: 2000, p. 690.

[4] G. Jervan, Z. Peng, R. Ubar. "Test Cost Minimization for Hybrid BIST," *IEEE Int. Symp. on Defect and Fault Tolerance in VLSI Systems*. Tokio, October 25-28, 2000, pp.283-291.

[5] A. Jutman, R. Ubar, "Design Error Diagnosis in Digital Circuits with Stuck-at Fault Model," *Journal of Microelectronics Reliability*. Pergamon Press, Vol. 40, No 2, 2000, pp.307-320.

[6] R. Ubar. "Dynamic Analysis of Digital Circuits with Multi-Valued Simulation," *Microelectronics Journal,* Elsevier Science Ltd., Vol. 29, No. 11, Nov. 1998, pp.821-826.

[7]   Turbo Tester home page URL: http://www.pld.ttu.ee/tt

[8]   REASON project home page: http://reason.imio.pw.edu.pl

[9]   Java applets home page: http://www.pld.ttu.ee/applets

[10]   Laboratory training URL: http://www.pld.ttu.ee/testing/labs

A PORTABLE DEMONSTRATION SET-UP FOR ANTENNAS AND PROPAGATION TEACHING

MOSIG J., ZURCHER J.-F., BRUEGGER S.
(LEMA, Swiss Federal Institute of Technology Lausanne, Switzerland),
http://itopwww.epfl.ch/LEMA; jf.zurcher@epfl.ch
Key terms
Antennas and propagation, teaching, demonstration setup

Abstract
A portable demonstration setup for antennas and propagation teaching has been studied and realized. It allows numerous "live" demonstrations: antenna directivity and gain, polarization effects (co- and cross-polarization with both linear and circular polarization), effect of reflections, as well as information transmission at X-band. The whole setup is contained in a lightweight suitcase and is completely autonomous thanks to battery power supply.

1.   Introduction

Teaching of microwave antennas and propagation is not an easy task, and a simple but effective demonstrator might help students in understanding some basic principles. This is the main reason of the realization presented here.

2.   Realization

The basic idea for the demonstrator was to develop a simple educative tool that would allow impressive demonstrations to students at microwave frequencies. It consists mainly of a low-power transmitter with modulation capability, a simple receiver, various antennas, printed grids for both linear and circular polarization, and a reflector. Each item is described in more details hereafter.

2.1   TRANSMITTER

The operation frequency was chosen in the popular X-band, due to price and dimensions considerations. For the transmitter, a commercial waveguide cavity Gunn diode oscillator with an output power of about 10 mW was chosen. It has the advantage of simplicity and ease of operation from batteries. Two types of modulation can be applied to this oscillator, which is driven by a MOS transistor:

- a simple 1 kHz square signal is used most of the time; this makes possible the realization of a simple receiver

A.M. Ionescu et al. (eds.), Microelectronics Education, 259–268.

- for demonstrations where music or other information has to be transmitted, PWM (Pulse Width Modulation) will be used. This technique is the simplest one giving good audio quality results when modulating a Gunn diode oscillator.

The power supply voltage of the transmitter is 12V, delivered by 8 AAA batteries.

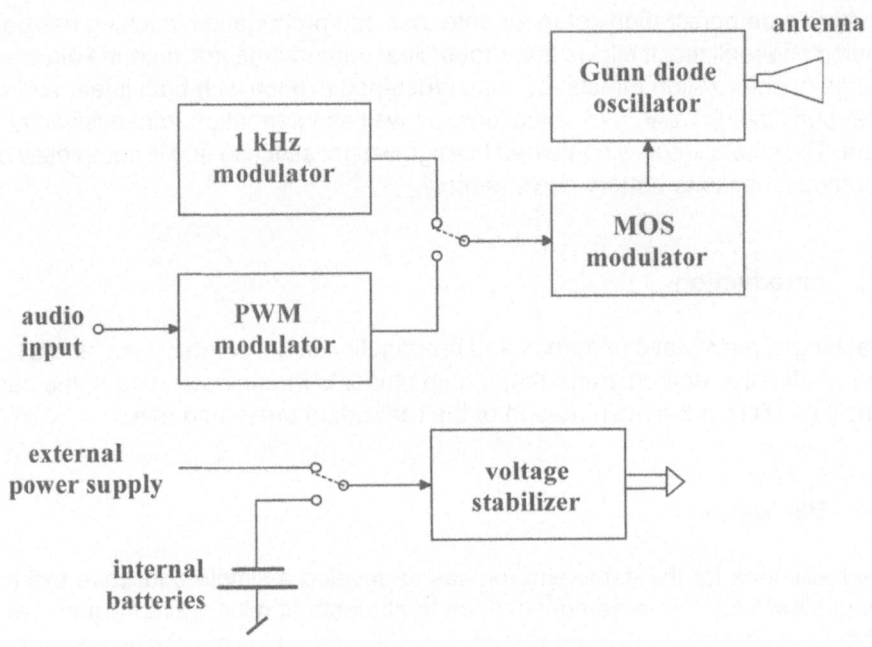

Fig. 1: schematic drawing of the transmitter

The complete transmitter (Figure 1) with all its electronic circuitry (voltage stabilization for the Gunn oscillator, AM and PWM modulators) and the batteries are enclosed in a plastic box measuring 190x110x70 mm. This box includes a support which will be used to mount the various polarization grids. The transmit antenna can be mounted directly on the waveguide flange protruding from the box. Figure 2 shows the complete transmitter.

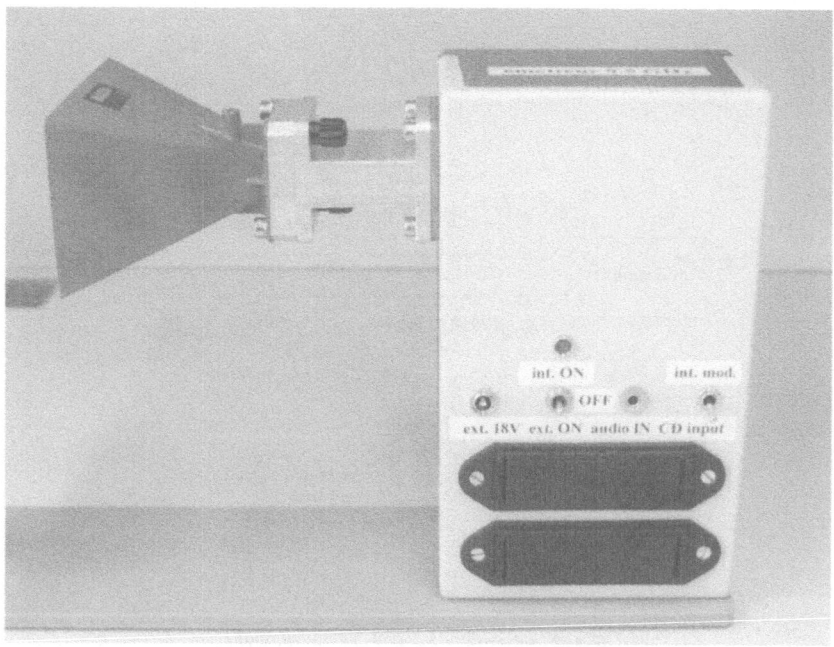

Fig. 2: external view of the transmitter with 15 dB horn antenna mounted

2.2   RECEIVER

The simplest way to demodulate a microwave signal is to use a classical waveguide diode detector followed by amplifier stages. The receiver (Figure 3) comprises an input preamplifier, followed by two paths:

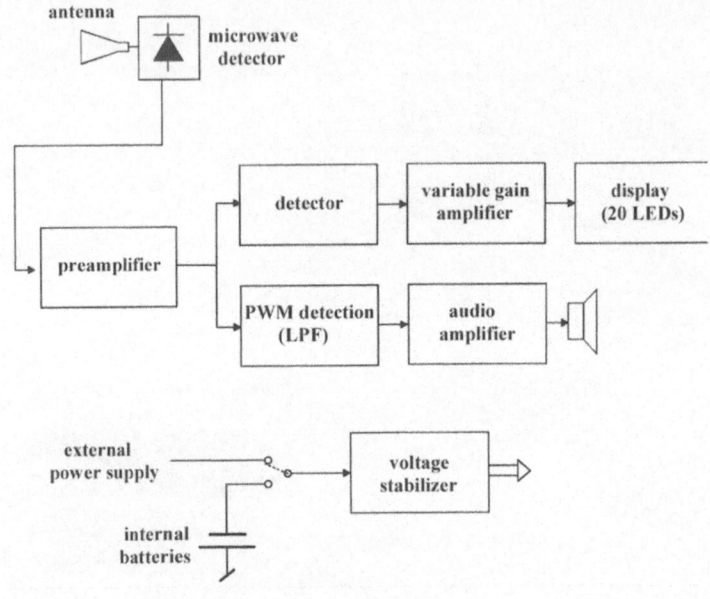

Fig. 3: schematic drawing of the receiver

- The "visual" path, comprising a variable gain amplifier, a detector and a display section made of two LED drivers and 20 high brightness red LEDs displaying the relative received level with a scale of about 1.5 dB/LED. So the dynamic range is about 30 dB.
- The "audio" path, made of a low-pass filter to extract the audio signal from the PWM signal, and an variable gain audio amplifier followed by a loudspeaker

Additional circuitry stabilizes the 12V delivered by the batteries (8 pieces identical to those used in the transmitter). The receiver is shown open in Figure 4. The case dimensions are 235x210x90 mm.

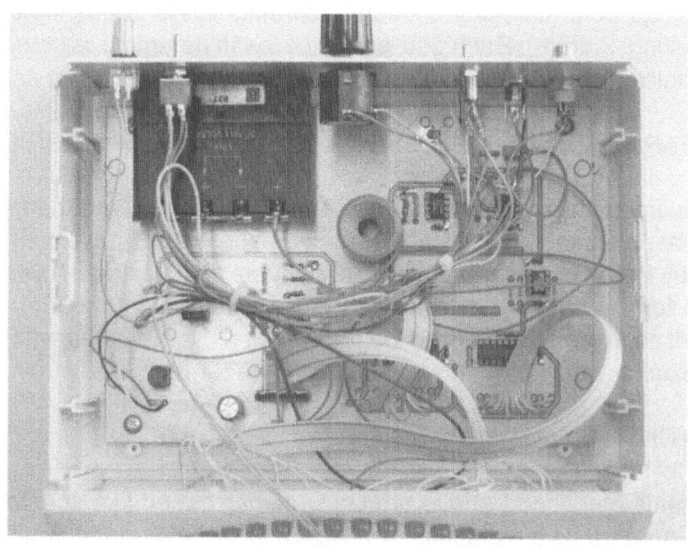

Fig. 4: internal view of the receiver (at the top the control panel, at the bottom a part of the 20 LEDs display)

A support to mount the receiving antenna can be placed on a shaft directly at the top of the plastic box containing the receiver. It can also be placed on a box containing spare batteries. This support includes a rotating part which allows the receiving antenna to be turned and locked at fixed angles by 45° steps to study the effect of co-, cross- and slanted polarization.

2.3   ANTENNAS

Various antennas are used for the demonstrations: two 15 dB gain horns, a 10 dB gain horn, as well as a specially-designed slot-coupled patch antenna with about 6 dB of gain. For additional demonstrations, a small parabolic antenna can also be used. All antennas are mounted on the waveguide flanges using screws.

2.4   REFLECTOR

The reflector consists simply of a 1.6mm thick FR4 printed circuit board completely metallized. This forms a rigid but lightweight reflecting plate measuring about 380x260 mm.

2.5   LINEAR POLARIZATION GRIDS

Two such grids were realized on FR-4 printed circuit material. They consist of a series of parallel printed lines etched on the FR-4 material. The width and spacing of the lines has been carefully chosen according to the frequency used (width: 0.5mm, spacing: 2.5mm). Each grid is provided with mounting structures allowing their positioning at 0°, 45° and 90° orientation.

2.6 LINEAR TO CIRCULAR POLARIZATION CONVERSION GRIDS

These grids are multi-layered structures designed after [1]. 4 meander-lines structures have been etched on a 0.1mm thick FR-4 substrate. These 4 structures are separated by Rohacell 51 foam layers of adequate thickness. All layers are glued together to form a rigid structure. Measured on the whole 8.2-12.4 GHz band, these polarization conversion grids convert a purely linear polarization into a circular polarization which presents an axial ratio better than 1 dB.

2.7 SUITCASE

All components mentioned above are conveniently stored in a specially adapted suitcase (Figure 5). Its dimensions are 460x370x190 mm, and the total weight is about 7.5 kg, including batteries (1 set inserted in the transmitter and receiver, and 1 spare set).

3. Antenna properties demonstrations

To demonstrate antenna properties, one can use the two different horn antennas (10 and 15 dB of gain) as well as the microstrip antenna. Starting with the large horn, the decrease in the received signal level can be clearly seen when one mounts the small 10 dB horn. Changing now to the microstrip antenna, the received signal goes down again. At the same time, when turning the receiving antenna, the beamwidth of the antenna can be clearly seen: one can for instance measure the rotation angle for a given decrease in signal received. The beamwidth depends on the gain of the antenna: the lower the antenna gain, the larger the beamwidth of the antenna.

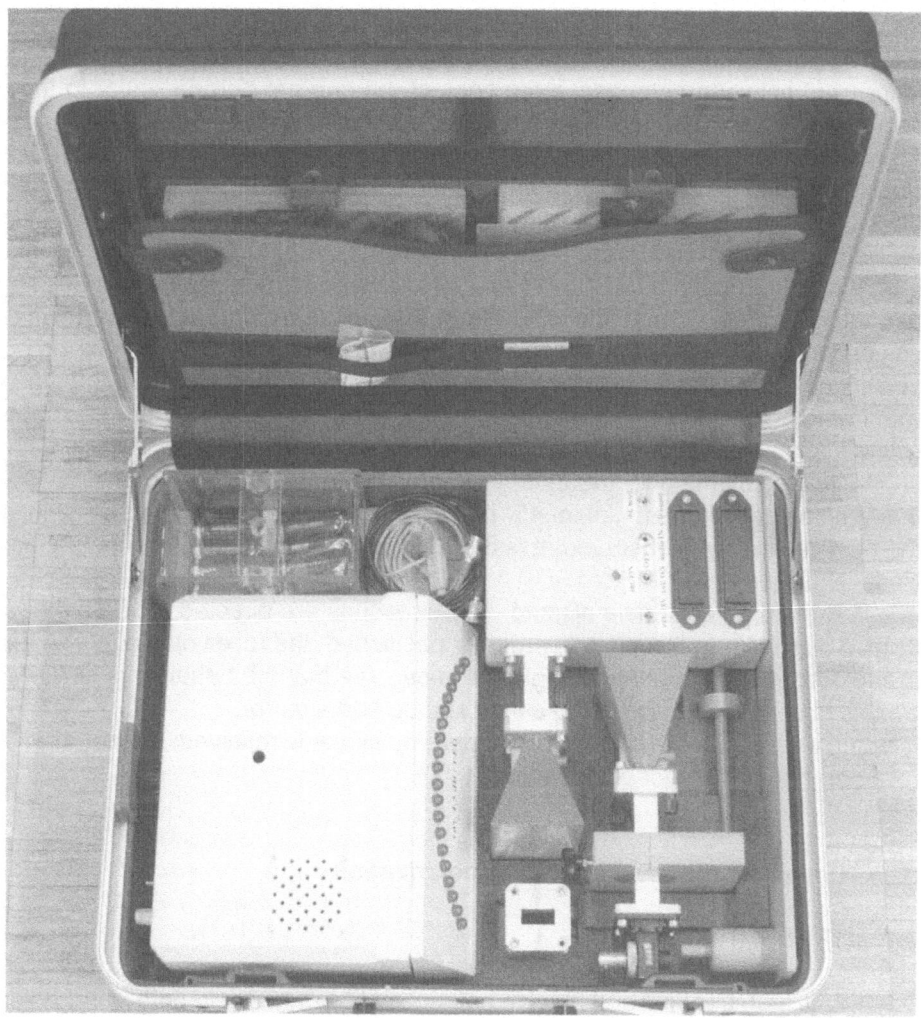

Fig. 5: suitcase with transmitter, receiver, antennas and supports, varius cables and polarization grids (at the top)

4.   Demonstrations with linear polarization

Using standard linear polarization (the "natural" polarization of the antennas used here), many demonstrations can be made:

4.1. crosspolarization: using the rotary support carrying the receiving antenna, one can turn the antenna by 45° steps. Turning the antenna to 45°, the received level decreases by 3 dB. By further turning to 90°, the received signal practically disappears (if everything were perfect, it should be zero).

4.2. The polarization being the same at transmit and receive sides, the linear grid is inserted between antennas; if the lines of the grid are perpendicular to the E vector, the received signal level remains practically the same, i.e. the grid is "transparent". On the other hand, if the lines of the grid are parallel to the E vector, the signal level decreases substantially, indicating that the grid reflects most of the signal. This can be explained by the fact that the lines of the grid short circuit the signal when parallel to the E vector.

4.3. The same experiment is repeated with orthogonal polarizations at transmit and receive sides. If the grid is oriented at 0° or 90°, no signal is received, which can be very easily explained (in each case, the grid "short circuits" one of the polarizations). If the grid is now placed at 45°, a signal 3 dB lower than the signal with both polarizations aligned appears. This can be explained by the decomposition of any polarization in two orthogonally-polarized signals, one of them being short-circuited by the grid, the other one reaching the receive antenna.

4.4. With a relative orientation of 45° between transmit and receive polarizations, the received signal is observed. It is 3 dB lower than it would be with both polarizations aligned.

4.5. With both polarizations aligned, both antennas are directed towards the reflective plate. When this plate is properly positioned, the received signal is maximized. Observing the relative angles, one can see that the behavior is similar to what is observed in optics using a light source and a mirror.

If both polarizations are oriented 90° apart, no signal is received, neither directly (see paragraph 4.1), nor via reflection.

5.   Demonstrations with circular polarization

5.1   TRANSMISSION

5.1.1. With both transmit and receive antennas aligned and with both linear polarizations being the same, the display is adjusted on the receiver. Then a polarization conversion grid is inserted near the transmitting antenna: the received level decreases by about 3 dB. This is due to the fact that a circularly-polarized wave is received with a linearly-polarized antenna.

A second, identical polarization grid is then placed near the receiving antenna (see Figure 6); depending on its orientation, the received level is either the same as it was

Fig. 6: setup ready for demonstration of circular polarization: receiver with 15 dB horn, polarization grid, polarization grid, transmitter, CD player (from left to right)

without the grids, or nearly zero. In the first case, we have two circular polarizations of the same type, in the second case the circular polarizations are of opposite sense of rotation.

5.1.2. When both grids are oriented so that both circular polarizations are the same (maximum of received signal), the transmitting antenna with its polarization grid is rotated. There is no change in the level of the received signal, showing an interesting property of circularly-polarized waves (application to satellites!).

5.2   REFLECTION

Placing both polarization grids near the transmitting and receiving antennas respectively, so that transmission is maximum (same sense or rotation of the circular polarization), the received level is observed. Then the transmitting and the receiving antennas are oriented (together with their respective polarization grids) towards the reflector. Trying to adjust the position and angle of the reflector gives no received signal.

Then, the orientation of one of the polarization grids is changed; in this case, a received signal appears when the reflector is properly positioned and oriented: this proves that the sense of rotation of the circular polarization is inversed by a reflection. If now the transmitter is directed directly to the receiver, no signal is received, because both polarizations have an inverse sense of rotation.

6.   Transmission of information

As mentioned earlier, the transmitter can be modulated (PWM modulation). Any sound source (for instance a portable CD player) can be connected to the transmitter. All what remains to do is to switch the transmitter from "internal" to "external" modulation. On the receiver, the audio amplifier has to be switched "ON". The audio transmission is of good quality. This is a good demonstration of a sound transmission via microwave link.

Conclusions

A lightweigh (~7.5 kg including suitcase), autonomous portable demonstration setup has been studied and realized. It allows numerous "live" demonstrations in the domain of antenna and propagation, and is an invaluable help for courses and demonstrations. Thanks to a careful energy management, it has an autonomy of about 20 hours with a set of standard batteries.

Due to the simple and clever design, it takes less than one minute to install the whole system to be ready for a demonstration.

References

[1]   L. Young, L. A. Robinon, C. A. Hacking, "Meander-Line Polarizer", IEEE Transactions on Antennas and Propagation, May 1973, pp. 376-378.

TOUCH-EE-TO-GRASP-AI

SALOMON R., KRUMPHOLZ B.
University of Rostock, Department of Computer Science and Electrical Engineering, 18051 Rostock, Germany

Bringing AI Fascination to Highschool

ABSTRACT
This paper reports on a summer school that was run for highschool students. In order to provide a fair overview of typical studies in electrical engineering (EE) this first summer school was offering several lectures in the field of new artificial intelligence and robotics. The topics were taught on various levels ranging from technical basics to high-level programming. An informal evaluation has indicated that the chosen topics were suitable for the summer school's goals and that the attendees might be interested in starting their studies in EE.

KEYWORDS:
New Artificial Intelligence, Education, Robotics, Summer School

1.   MOTIVATION

Mecklenburg-Vorpommern is one of Germany's sixteen states. It faces severe economical problems: the economy is low, industrial companies are leaving or closing down; the unemployment rate has surpassed the rate of 20% [5]. The state also faces social problems, which are partly due to migrations within Germany. About 50% of the migrants are in the age of 20-35 [5]. The lack of adequate job opportunities in this region also affects the students' willingness to do their studies at the University of Rostock. The gap between the number of students in the engineering departments and the open positions in the field of EE and computer science (CS) [8] in Germany is, on the other hand, too large. Informal inquiries have indicated the following serious reasons. The awareness of engineering is very low at the regional highschools and many highschool students have no or a rather fuzzy idea about the profession "engineer". Highschool students simply assume that programming and CS will not be part of the EE studies. Traditional studies in EE are thought of being very hard and theoretical. Teachers rather send interested students to physics and math. In light of this situation it might come to a surprise that highschool students express significant interests in EE at various occasions. For example, the Institute of Applied Microelectronics and CS at the University of Rostock has been running a robot race competition, called SPURT [7]. SPURT runs different leagues in order to account for various aspects. During both the race event and its preparation many attendees have expressed strong interests in robotics, programming, and EE.

269

A.M. Ionescu et al. (eds.), Microelectronics Education, 269–273.
© 2004 *Kluwer Academic Publishers.*

Because of the described development, i.e. increasing interests in various EE aspects and the gap between the open positions and the number of students, the institute did informal interviews indicating that many schools cannot provide the required resources such as teachers, equipment, course material, etc. Based on this situation, the institute has decided to run a summer school for highschool students. It should be mentioned here that opposed to the other institute's activities such as SPURT and the Girls' Day [13], this summer school is rather "professional" and offers a wide area of topics as well as general information.

2.   The Summer School

2.1   GOALS

The main goal of the organizers is to attract highschool students to the studies of EE. That is why the summer school should give the attendees a fair overview of typical studies; a bias within this field towards the organizing institute's interests might be ok, though. With respect to choosing the class material, it should be kept in mind that presenting both a broad overview and some detailed material might be conflicting with the duration of only one week. Furthermore, such an event should be fun for the students because the word of mouth between colleagues is the best advertisement for the future.

The attendees' goals might be different. They voluntarily spend part of their summer break in order to do some extra studies (without immediate benefits). Therefore, the content should match the student's interests, the material should be introductory but also challenging and it should allow plenty of practical work. The summer school also indicated that the students expect it to inform about the studies in EE and information technology and everyday campus life. Furthermore, most attendees would like to extend their knowledge. When asked, the students expressed that they did not want to just repeat regular school material but that they expect new topics.

2.2   COURSE CONTENT

New artificial intelligence (AI) [2] aims at understanding the underlying mechanisms of natural intelligence by building physical robots that autonomously operate (without any human control) in dynamically changing environments. Because of this combination, new AI matches the aforementioned goals quite well and thus, seems to be an ideal focus of the summer school presented here. The targeted audience are students in their last three years of highschool in Rostock and its vicinity. It can thus not be assumed that all attendees do have the same previous knowledge and that they do have practical experiences with soldering, measuring, etc. The summer school's [6] class material included:

Fundamentals: handling of components, soldering, measuring, circuits
Background: "ice breaker" / how to handle all sorts of equipment
SPURT robots: building simple physical SPURT-robots
Background: how things operate / why they actually work
MindStorms: basic programming skills by using LEGO [9] and NQC [12]
Background: higher-level techniques / programming professional black boxes
Mobile robots: Braitenberg vehicles [1], Khepera Simulator [3, 4]
Background: agent-environment interaction / simulation experiments
Bluetooth: programming ad hoc networking / Visual Basic
Background: up-to-date communication techniques / embodied AI
Excursion: institutes of laboratory automation and bio-chemical analytics
Background: modern working situations / interdisciplinary form

3.   First Experiences

3.1   FORMAT AND ATTENDEES

This summer school was run during the first summer break's week of the regional highschools with eight hours per day. Afterwards, the students had the opportunity to do some work for an additional hour if desired. In order to guarantee an intensive learning atmosphere the number of attendees was limited to 16; due to some unexpected walk-ins, 20 students were participating. The practical work was done in small teams, mostly consisting of two students. It was interesting to note that within these small groups no competition has emerged. It furthermore could be noted that none of the teams were competing with any other. It could be that the lack of any exams and grades was a reason for this cooperative atmosphere. During the lectures, it could be observed that the students do not like "long-winded'" theories but prefer a mix of short (theoretical) lectures and practical exercises in which they could experiment with the material and apply their knowledge. The attendees were from nine highschools and of ages between 15 and 22; five participants were holding their highschool diploma. Two or three graduate students were supervising and helping the students. In order to allow for intensive practical work, the institute provided eight fully equipped work places.

3.2   EVALUATION

At the end, the organizers were doing an informal evaluation, where the attendees had to complete a simple, one-page evaluation form. The majority of the attendees graded the summer school with a good, and a third even with a very good. They also thought that the material was presented on an adequate level. According to the evaluation forms, the attendees appreciated the total number of attendees and liked to work in couples. Furthermore, they did not complain about the attendees' diversity (in terms of age and knowledge). Asked about the topics, the attendees mentioned that they liked putting together these robots the most. They apparently appreciated to experience the various levels i.e., from pure hardware to high-level programming, on which they can construct robot controllers. The lecturers, on the other side also did an informal evaluation. They all thought that the students possessed the required pre-knowledge in order to follow the presentation. Furthermore, they all had a very positive impression about the summer school: very informal format and atmosphere. None of the lecturer ever noticed any kind of "bad competition"; rather, the students were helping each other. It seemed as they all wanted having fun and being successful. This observation is in quite contrast to the typical learning situation in schools and universities. Overall, the attendees and the lecturer had a very positive impression.

Conclusions

This paper has reported on a summer school to attract highschool students to the studies on EE. An informal evaluation has indicated that the attendees liked the summer school very much. They particularly liked building robots and visiting other institutes since they could grasp an idea about every-day work. Furthermore, the students were exposed to some major ideas and concepts of new artificial intelligence. The informal evaluation indicated that highschool students prefer practical work. The lecturers as well liked the summer school and are more than willing to repeat such efforts. A recent study [14], also known as PISA, indicated some problems of the German education system and has initiated discussions about possible changes [10, 11]. Major topics include new contents and educational standards. Another important goal is to better interconnect highschools and universities. Since the chosen course content, mainly new artificial intelligence, contributes to the education in the technical and engineering fields, the summer school presented in this paper is well situated in these nation-wide activities.

References

[1] V. Braitenberg, VEHICLES, Experiments in synthetic psychology, MIT Press, Cambridge, MA.1984

[2] R. A. Brooks, Intelligence without representation. Artificial Intelligence, 47, 1991, 139-159.

[3] Khepera robots: http://www.k-team.com/

[4] Khepera simulator: http://diwww.epfl.ch/lami/team/michel/khep-sim/

[5] Statistisches Landesamt M-V: http://www.statistik-mv.de/

[6] Summer school: http://www.e-technik.uni-rostock.de/sommerschule/

[7] SPURT: http://spurt.uni-rostock.de/

[8]   Innovations-Report: http://www.innovations-report.de/

[9]   LEGO: http://www.lego9.com/

[10] Technique Didactic Forum: Educational Standards and Quality Assurance in Schools, Universities and Seminars, 2003, Berlin, http://www.ph-freiburg.de/tehawi/wwwdgtb/

[11] http://www.lernnetz-sh.de/pisa/ and http://www.learn-line.nrw.de/

[12] Not Quite C: http://bricxcc.sourceforge.net/nqc/

[13] Girls' Day: http://www.girlsday.de/

[14] Programme for International Student Assessment (PISA): http://www.pisa.oecd.org/

POSTER PRESENTATIONS

PROGRAMMABLE LOGIC AND WEBLABS

ZUBIA J.-G.
Faculty of Engineering. University of Deusto. Apdo. 1. 48080 Bilbao. Spain.
zubia@eside.deusto.es

1. INTRODUCTION

One of the goals of modern universities is to decentralize part of its activities, giving students more freedom to organize their own schedules. In this sense, the European Union now faces a new challenge: the reforms introduced by the Bologna Convention. In this new education framework, students will have more freedom to organize their time, which will result in loosely supervised schedules that will make the running of labs complicated.

In this new scenario, the idea of a WebLab is very appealing: take the lab out of its traditional setting and offer it through the Internet. WebLabs have been growing spectacularly during the past decade, and many important and different centers use them (MIT in the USA [1], Universities of Deusto [2] and Valencia [3] in Spain, Centro Tecnológico Pereria in Colombia [4], University of Siena in Italy [5], and many distributed around the world).

This paper reports on our department's experience with a WebLab in the field of Programmable Logic with CPLD and FPGA. First, we describe what a WebLab is and what its advantages are. Next, we describe "PLD WebLab", the weblab developed at the University of Deusto. Finally, the paper concludes by reviewing the possible improvements to the WebLab and future lines of work.

2. WHAT IS A WEBLAB?

A WebLab allows a lab's services to be offered through a computerized medium (the Internet). In other words, a WebLab allows students to carry out their practical lab exercises outside the lab.

There are two kinds of WebLab: Virtual and Remote. The *Virtual Laboratories* are simulation programs with an extensive use of multimedia (image, simulation, texts, etc.), so the student uses virtual instrumentation: oscilloscopes, wave generators, motors, etc. Virtual Laboratories are not hardware, they are computer simulations.

The latest trend in WebLabs is the *Remote WebLab* in which a student accesses hardware equipment through a WWW interface. Through that interface, the student can program, control, and observe the evolution of his actions in realtime through a webcam or similar device. Thus, the student can work with hardware from his home, or any other place, in a realistic fashion, without the layer of

A.M. Ionescu et al. (eds.), Microelectronics Education, 277–281.
© 2004 *Kluwer Academic Publishers.*

abstraction that simulations add. The WebLab presented in this paper is a Remote WebLab.

The design and use of a Remote WebLab has several advantages:

[2] Higher availability of lab equipment (365 days per year, 24 hours per day).

- Lab organization. The labs don't have to be physically open at all
  times, since it is enough to keep the WebLabs up and running.
- Efficient schedules. Using WebLabs, both students and teachers can
  organize their time with more ease, creating efficient schedules.
- Independent work. WebLabs encourage independent work and study,
  one of the cornerstones of the new european education system.
- Community involvement. WebLabs allow all the community (both
  academic and non-academic) to benefit from the labs.
- e-Courses. WebLabs are an important tool to organize e-Courses (i.e.
  courses taught remotely, either through the Internet or
  videoconference).

Besides these academic advantages, there are broader implications:

- Handicapped students. WebLabs allow people with physical or mobility
  impairments to use sophisticated lab equipment through the Internet
  using specially designed software.
- Third world. WebLabs can bring technology closer to Third World
  countries.
- Access to remote resources. WebLabs can be used at universities to
  access equipment that, due to its cost or complexity, is only available in
  certain companies or research centers.

3.   WEBLAB PLD

The Department of Computer Architecture of the University of Deusto has designed a Remote WebLab oriented toward Programmable Logic: *The PLD
WebLab*. This hardware/software environment allows a student/user to remotely
access a Xilinx CPLD XC9572 device, program it, try out different input combinations, and observe the outputs through a webcam.

One of the pillars of our design strategy has been to use simple software and
hardware. The goal of this project is not only the development of the WebLab itself, but also to show that WebLabs are not necessarily complex environments
for expert users.

3.1   WEBLAB STRUCTURE

The PLD WebLab consists of a TCP/IP-based client/server architecture. The cli-
ent and server programs are written in C, and the underlying hardware is a CPLD
XC9572 device, a PIC-based communications board, application boards (motors,
LEDs, etc.), and a webcam (see Figure 1).

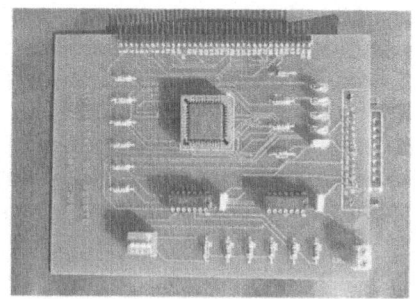 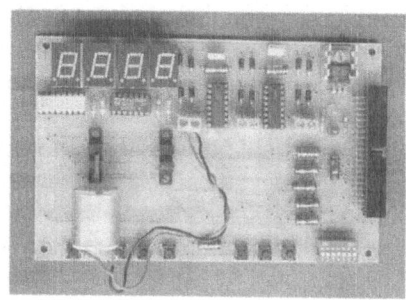

Fig. 1: CPLD and motors-based application boards

Once the server is up and running, any client can start a connection and take control of the CPLD device. A complete work session would consist of the following steps:

- The user writes a program in VHDL, ABEL, etc., simulates it, and when he believes the program is correct, generates the corresponding JEDEC file.
- The user starts a connection with the server.
- The user sends the JEDEC file to the server.
- The server programs the CPLD with the JEDEC file (through the parallel port and using the JTAG standard).
- The student controls the inputs of the application using the client software, the PIC and the RS-232 standard.
- The CPLD controls the motors, LEDs, etc using the programmed algorithm.
- The student sees the changes through the webcam, which is constantly monitoring the CPLD and any attached devices.
- At this point, the user can try out other input signals, or close the connection with the server.

The client/server softwaret is a simple C program, and does not have a GUI. The software allows the user to program a JEDEC file on the remote CPLD device, try out different input signals (switches, push buttons, and the clock signal), change his password, and see how long he has been connected to the server. While he is controlling the CPLD device, the user can see the result of his actions in the application board through the webcam, which is made available through a web address.

3.2   SPECIFICATIONS AND LIMITATIONS

The PLD WebLab has the following specifications: custom made client/server software, hardware (CPLD, communication, and additional devices), server with

Pentium microprocessor and 32 bit operating system (Windows 2000), DSL Internet access, serial and parallel ports, basic webcam, and a website where the webcam images can be accessed. The limitations of the WebLab are mainly due to slow Internet access (which limits the effectiveness of the webcam), problems with error recovery, and the need to open ports 8080, 2003, 20, and 21 in the university's firewall.

The PLD WebLab has been thoroughly tested, and it will soon be used by students of a course of Programmable Logic in the engineering faculty. We hope that during the 2004/05 academic year, PLD WebLab will be available to students from other universities around the world.

CONCLUSIONS

Thanks to our work with PLD WebLab, we can conclude that a functional WebLab can be developed with little effort for practically any electronic equipment, which improves the performance of hardware labs by maximizing their availability, allows courses to be held without requiring the student's physical presence, and opens labs to society and its needs.

The work we have presented is a useful and simple WebLab prototype. Our challenge is now to develop a more professional one, which will mean improving the existing prototype in the following ways:

- Conecting several CPLD and FPGA-based boards to the server, no just one.
- Rewrite the client/server software in Java or a similar language which will allow the development of a more elaborate GUI.
- Design a website that provides easy access to all of the WebLab's services.
- Improve security issues in the server.
- Create usage policies for the hardware so its availability can be improved.
- Allow users to recover remotely from errors.
- Integrate the WebLab in a multimedia environment.

As for future lines of work, we are currently concerned with the fact that the design of the WebLab is the responsability of individual departments, or even individual laboratories. Each department ends up having its own different WebLab, with different interfaces, usage policies, etc. Since the are no uniform criteria for WebLab design, the student can be confused by all the different WebLabs. In our opinion, the design of a WebLab should be carried out by the faculty or the university, so each of the individual WebLabs would have a similar architecture, look & feel, and interface. Using this approach, each department (and its staff) would specify the equipment and practical exercises they want to make available through a WebLab, and a new entity, the "WebLab Service", would be in charge of its design and maintainance. This, in fact, doesn't clash with the current approach to Internet and Web Services in universities: it is normal to have a university-wide service in charge of web design and maintainance that responds to each department's

needs. The real challenge is to define an architecture and all necessary procedures to effectively design WebLabs homogeneously within the same faculty or
university.

Acknowledgment

PLD WebLab is part of the project LOGBOT supported by the Regional Goverment of the Basque Country (Spain): Department of Industry, Commerce and
Tourism, SAIOTEK 2002, OD02UD05.

References

[1]   Jesús A. Del Alamo, MIT Microelectronics Weblab. Marzo, 27, 2001. http://
       web.mit.edu

[2]   Larrauri, I.; García, J. and Kahoraho, E. "Integration of WebLab Systems in
       Engineering Studies". Proceedings ICEE International Conference on
       Engineering Education. ISBN: 84-600-9918-0, 5 pp in CD, Valencia,
       (Spain), 2003.

[3]   Rodrigo, V.M; Bataller, F.M.; Baquero, M. and Valero, A. "Virtual
       Laboratories in Electronic Engineering Education". Proceedings ICEE
       International Conference on Engineering Education. ISBN: 84-600-9918-0,
       5 pp in CD, Valencia, (Spain), 2003.

[4]   Pérez M. et al. "Laboratorios de acceso remoto. Un nuevo concepto en los
       procesos de Enseñanza-Aprendizaje".
       http://digital.ni.com/worldwide/latam.nsf/web/all/
       F54369A0EC8C0B4486256B5F006565A9

[5]   Casini, M.; Prattichizzo, D. y Vicino, A. "e-Learning by Remote
       Laboratories: a new tool for control education" . The 6th IFAc Conference on
       Advances in Control Education, Finland, 2003

TEACHING THE SYSTEMATIC ANALYSIS OF ELECTRONIC DEVICES WITH ANIMATED PRESENTATIONS

Demonstration of methodically solving the semiconductor device equations
SCHROEDER D.
TU Hamburg-Harburg, Microelectronics, D - 21071 Hamburg, Germany

1.   INTRODUCTION

The analysis of semiconductor electronic devices is a difficult task. The student, being confronted with the subject for the first time, has a hard time to follow the reasoning and to comprehend completely the approach because of various equations involved, different approximations called up, as well as "insider tricks" and "shortcuts". The problem is particularly severe with respect to boundary conditions. While the basic set of semiconductor equations is displayed prominently in the textbooks (for example [1], [2]), the same cannot be said for the boundary conditions. Often, they are introduced in an adhoc or even implicit manner. (A comprehensive collection of boundary and interface conditions can be found in [3].) Also the sequential nature of presentation makes it difficult to maintain the overall picture, and the student looses track of the underlying methodology.

As a conclusion, the requirement for demonstrating the systematic analysis of semiconductor devices in the classroom is identified. The method of teaching should include a complete specification of the device, the equations to be solved, the boundary conditions to be invoked, the approximations being made, and a clear illustration of the solution sequence.

It is the intention of this paper to introduce such a method of teaching the systematic analysis of semiconductor devices with the help of animated presentations. The method has been used by the author since several years in his lectures on "Electronic Devices" as part of the International Master's Program "Microelectronics & Microsystems " at TU Hamburg-Harburg.

2.   CONCEPT

The basic idea of the method is to maintain an animated display on the classroom screen where the solution of the device problem unfolds. The device geometry as well as the equations, conditions and solutions, assigned to their corresponding locations, are visible all the time. This helps the student to be aware of the current state of the solution including the solutions achieved so far, the assumptions being made, and the equations still waiting to be solved. So the student is kept on track of the calculation at any time.

At the beginning, the problem is established by showing the device geometry, indicating regions of different materials (semiconductor, insulator, metal) and

A.M. Ionescu et al. (eds.), Microelectronics Education, 283–288.

different doping concentrations. After that, the equations to be solved (Poisson's equation, continuity equations) are inserted into the respective regions, and corresponding boundary conditions appear next to the boundaries.

In the next step, the regions of interest can be further subdivided into subregions if necessary, thus preparing different approximations or solution methods in different subregions. Of course, corresponding conditions at the interfaces between subregions must be set up, and equations specifying the locations of the interfaces have to be added.

Now, the assumptions to be used are introduced in the respective subregions. After that, the solution of the problem unfolds by successively replacing the equations by their solutions and deleting conditions already satisfied.

3.   EXAMPLE

The method is illustrated by the example of a pn-diode in equilibrium. Figure 1 shows the statement of the problem, i.e. the 1D-diode with two regions of constant donor and acceptor doping and ideal contacts at both ends. Since equilibrium holds, the current continuity equations are satisfied by constant quasi-Fermi potentials ϕ_n and ϕ_p, which are zero. Thus, the only equation to be solved is Poisson's equation, as indicated in the middle of the picture. Neutrality boundary conditions are shown at the ohmic contacts. Figure 2 displays the screen after the depletion and quasi-neutrality approximations have been introduced. The device is now divided into three subregions where the respective assumptions hold. Poisson's equation as well as the boundary conditions have changed and reflect the approximations made. Two interface conditions each have appeared at the edges of the depletion region, ensuring continuity of the potential and the electric field. Two additional conditions specifying the concentrations at x_n and x_p are introduced, which determine the positions of the depletion region edges. Figure 3 shows a snapshot during the solution progress. Poisson's equations in the quasi-neutral regions have already been solved and were replaced by their solutions. The boundary conditions used for this have disappeared, ensuring that only equations still to be solved are visible. Figure 4 represents the final display. Poisson's equation in the depletion region is now solved too and turned into the solution. The integration constants C and D as well as the values for x_n and x_p have been put forward from the 4 interface conditions left in Figure 3.

REFERENCES

[1] Y. Taur, T.H. Ning, "Fundamentals of Modern VLSI Devices", Cambridge University Press, 1998.

[2] H.-G. Unger, W. Schultz, G. Weinhausen, "Elektronische Bauelemente und Netzwerke I", Vieweg, 1979.

[3]   D. Schroeder, "Modelling of Interface Carrier Transport for Device Simulation", Springer, 1994.

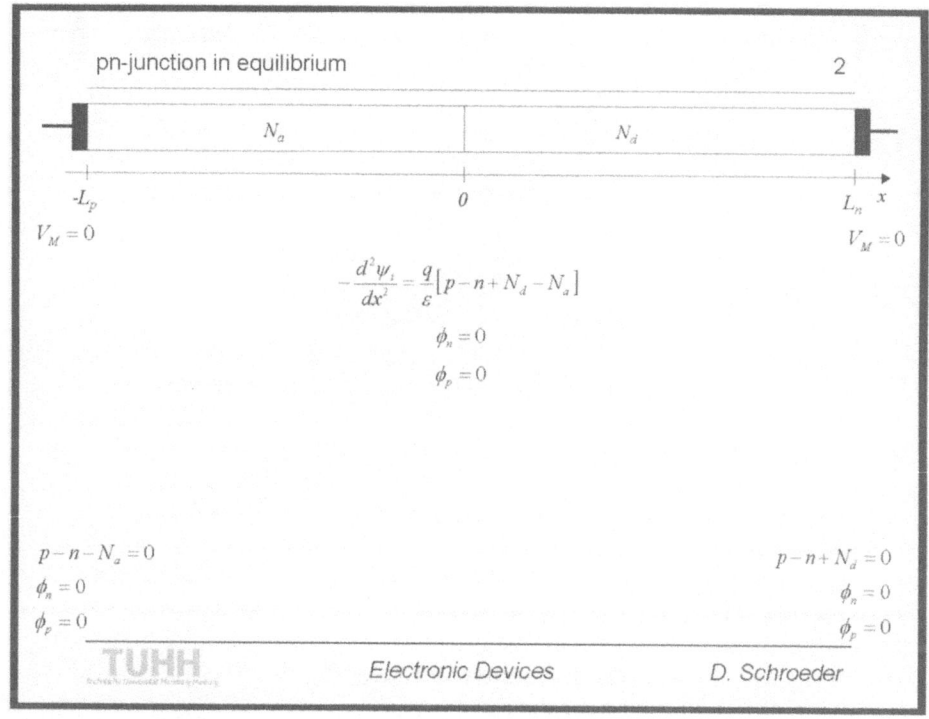

Fig. 1: Statement of the problem

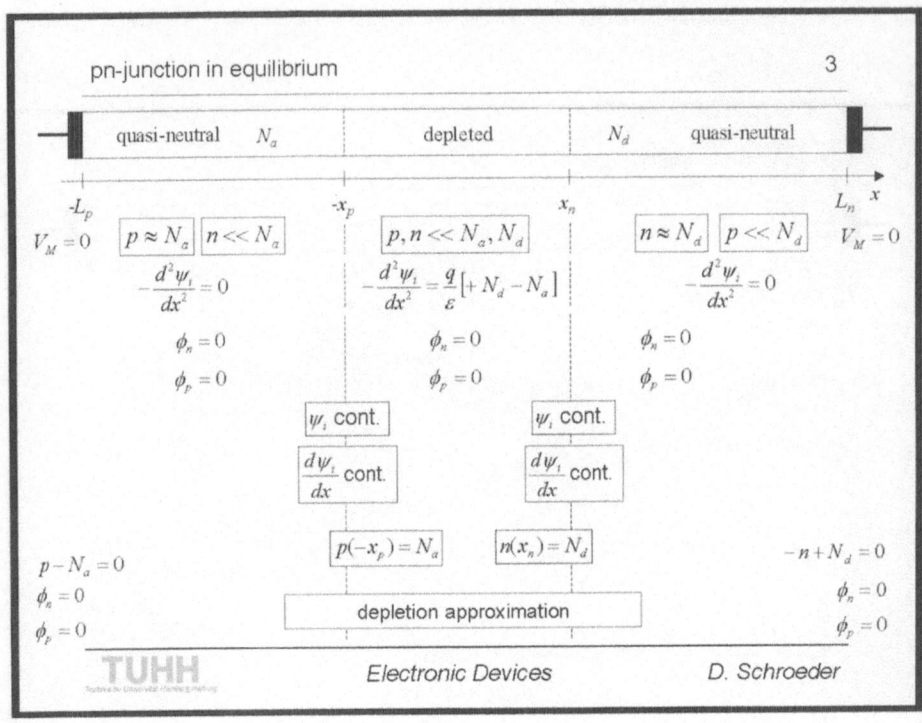

Fig. 2: Approximations introduced

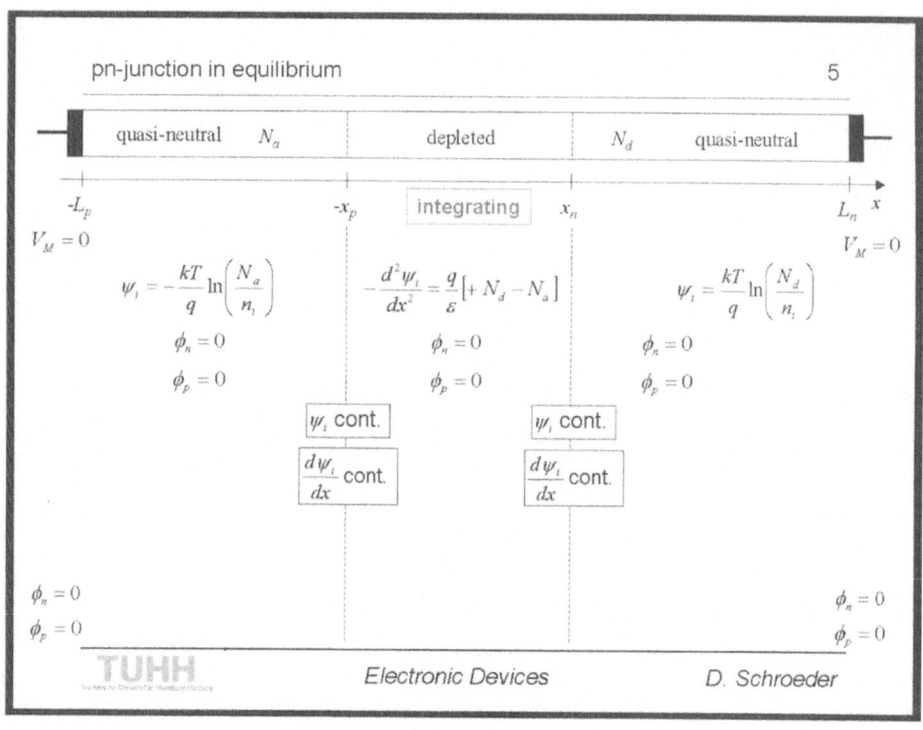

Fig. 3: Solving the equations

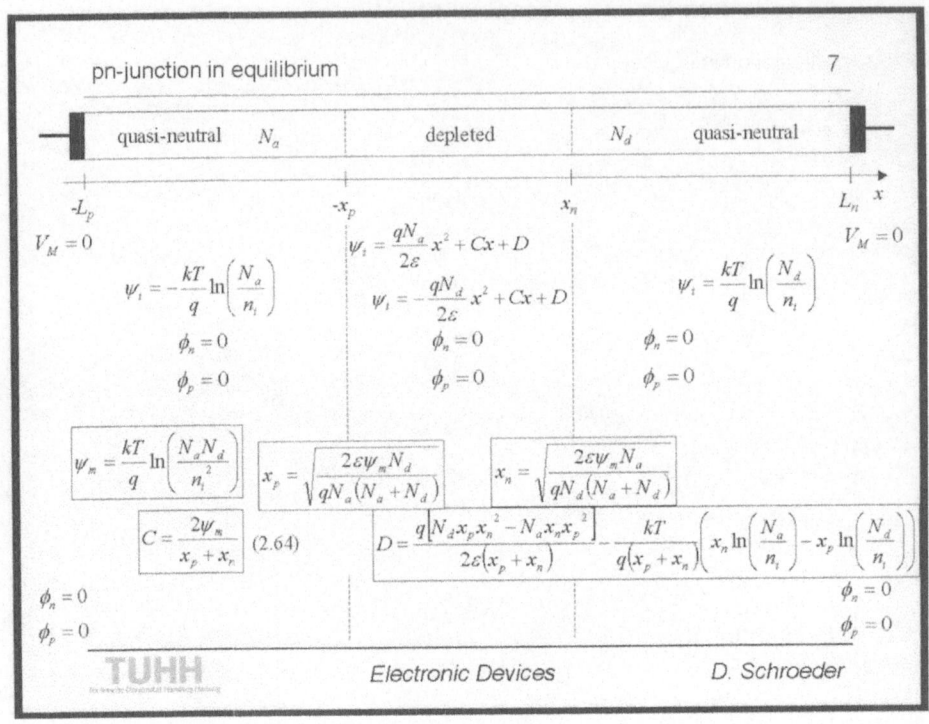

Fig. 4: The final solution

SOFTWARE FOR ANALYSIS AND DESIGN OF DIGITAL SYSTEMS IN EDUCATION: A COMPARISON BETWEEN BOOLE-DEUSTO AND LOGICAID

ZUBIA J.-G.
Department of Computer Architecture. Faculty of Engineering. University of Deusto. Apdo. 1. 48080 Bilbao. Spain. zubia@eside.deusto.es
SOTOMAYOR BASILIO B.
Department of Software Engineering. Faculty of Engineering. University of Deusto. Apdo. 1. 48080 Bilbao. Spain. bsotomay@eside.deusto.es

1.   INTRODUCTION

Software is an important teaching tool in introductory digital electronics courses. Many professional packages suit the requirements of these courses, such as Or-CAD, ISE (Xilinx), ElectronicsWorkBench, and MaxPlus II (Altera). However, while these packages are versatile, powerful, and used in the workplaces where students will end up in, they lack ease-of-use and the didactic approach which is needed by both students and teachers in introductory courses. Furthermore, professional packages focus mainly on the final results of a problem, instead of focusing on the methodology used to solve that problem.

The need arises for an education-oriented software package that can allow a student to perform the analysis and design of digital systems in a methodology-driven manner. These packages should not be a substitute for professional packages; they should be a complement which can ease the student's first steps in digital electronics. BOOLE-DEUSTO [1] (developed by the University of Deusto, Bilbao, Spain) and LogicAid [2] (developed by the University of Texas at Austin, USA) are two such software packages. This paper is divided into three parts: a short introduction to BOOLE-DEUSTO, a comparison between BOOLE-DEUSTO and LogicAid, and the conclusions of this comparison.

2.   BOOLE-DEUSTO

BOOLE-DEUSTO can carry out analysis and design of bit-level combinational and sequential circuits. The underlying philosophy of BOOLE-DEUSTO is to act as a 'boolean calculator' which can help the student perform digital electronics exercises, the same way a traditional calculator helps high school students perform math exercises. However, since BOOLE-DEUSTO is designed as a learning aid, and focused on methodology and not results, when a student uses it to solve an exercise, he will not only receive a result (which is what traditional calculator do), he will also be able to see the process used to arrive at that result.

A.M. Ionescu et al. (eds.), Microelectronics Education, 289–293.
© 2004 *Kluwer Academic Publishers.*

Table 1 summarizes BOOLE-DEUSTO's main features and methods. Figure 1 shows screenshots of BOOLE-DEUSTO's most popular features (the Veitch-Karnaugh module and the FSM interactive simulation).

Combinational Circuit	Finite State Machines
Boolean Expression	Moore-Mealy's Diagrams
Truth Table	FSM verification
Veitch-Karnaugh Diagrams	State Minimization
Canonical Forms	Moore -> Mealy Conversion
Minimized Expressions	Tables and Minimized Expressions
NAND/NOR Expressions	J-K and D Circuit Logic
Circuit Logic	Interactive and batch simulations
Code Generation: VHDL, OrCAD-PLD, ABEL, JEDEC, etc.	Code Generation: VHDL, OrCAD-PLD, ABEL, JEDEC, etc.
Program-wide features	
Save and load systems to/from disk	
Print systems in their various representations	
Associate text with a system	

Table 1. BOOLE-DEUSTO Features

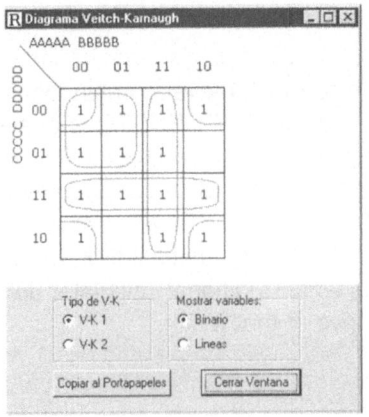 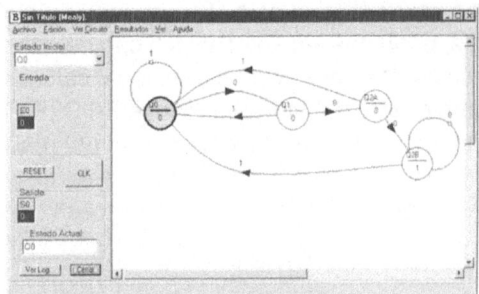

Fig. 1: Veitch-Karnaugh module and interactive FSM simulation

3.   COMPARISON BETWEEN BOOLE-DEUSTO AND LOGICAID

LogicAid, developed by the University of Texas at Austin, is another well- known software package which is suited for its use in digital electronics courses. Both BOOLE-DEUSTO and LogicAid are high quality teaching aids, yet they do not provide the same functionality. The goal of this paper is to analyze their differences between them and see how each could benefit from the other's features.

3.1   SUITED FOR DIGITAL ELECTRONICS EDUCATION

Both software packages have been designed from the start to be used mainly in a classroom setting, and therefore are well suited as an aid for digital electronics courses. Both programs feature a user-friendly interface which is easy to work with.

3.2   GENERAL STRATEGY

One of the primary concerns when designing BOOLE-DEUSTO was that the user should be able to use different representation methods (see 3.3) interchangeably with ease, enhancing the notion that several representations (a truth table, a circuit diagram, a Veitch-Karnaugh map) can all represent the same system. In effect, the user can input a system using a boolean expression, and then instantly see how that system would be represented with a VK map, in canonical form, etc. As mentioned previously, this makes BOOLE-DEUSTO act as a 'boolean calculator', where the user can input the description of a system and then experiment with it, the same way a high school student can experiment with all the features of a scientific calculator.

On the other hand, although LogicAid does allow the use of different representation forms, passing from one form to another one is cumbersome (the user has to save the system to a file and reopen it from the other representation form). The user generally has to stick to one representation method at a time, which makes the learning process too guided and not very open to experimentation.

3.3   FORMS OF REPRESENTATION

BOOLE-DEUSTO and LogicAid both allow the user to work with the most common forms of representation used in digital electronics (truth tables, Veitch-Karnaugh maps, canonical forms, etc.) However, BOOLE-DEUSTO has a more complete and versatile set of representation forms. The following are some noteworthy features of BOOLE-DEUSTO which cannot be found in LogicAid:

- Input a digital system in the form of any boolean expression, using any combination of AND, OR, XOR, NOT and parenthesis operators.
- The Veitch-Karnaugh module can also represent any system (LogicAid can only show systems with up to 5 variables).

- Circuit diagrams can be generated from the system description (both combinational and sequential)
- Sequential systems can be simulated graphically (using the FSM diagram) once they have been specified.

3.4   INTERNAL ALGORITHMS

LogicAid features much better internal algorithms than BOOLE-DEUSTO. While the latter only offers a simple (yet exact) single-output Q-M variant for boolean minimization, LogicAid offers several minimization methods, including the well-known Espresso method, which can perform multiple-output minimization.

3.5   EXPORT FORMATS

Both programs export combinational and sequential systems to well known formats. However, LogicAid currently only exports systems to JEDEC files, while BOOLE-DEUSTO can also generate OrCAD-PLD and VHDL code.

3.6   BIT-LEVEL AND WORD-LEVEL

Neither BOOLE-DEUSTO nor LogicAid work at word-level, since they can only perform bit-level analysis and design. However, LogicAid includes a partner program called SimUaid that allows the user to perform word-level analysis and design by directly editing and simulating logic circuits.

Feature	BOOLE-DEUSTO	LogicAid
General learning strategy	Boolean calculator	Guided learning
Representation methods	Many	Few
Internal algorithms	Normal	Excellent
Export formats	JEDEC, PLD, VHDL	JEDEC
Level of operation	Bit-level	Bit-level and word-level (SimUaid)

Table 2. Summary of comparison between BOOLE-DEUSTO and LogicAid.

CONCLUSIONS AND FUTURE LINES OF WORK
BOOLE-DEUSTO and LogicAid are the two leading software packages for digital electronics education. They are both very high quality products, each with a set of distinct features that makes them appealing to students and teachers.
BOOLE-DEUSTO is still being actively developed in the University of Deusto, and the two main lines of work in the future involve learning from what LogicAid offers

that BOOLE does not: improving the internal algorithms and supplying a word-level analysis and design tool.

Acknowledgment

BOOLE-DEUSTO is part of the project LOGBOT supported by the Regional Goverment of the Basque Country (Spain): Department of Industry, Commerce and Tourism, SAIOTEK 2002, OD02UD05.

References

[1]   García Zubía, J. "Problemas resueltos de electrónica digital". Thomson-Paraninfo, Madrid, España, 2003.

[2]   Roth, Charles H. "User's guide and reference manual for LogicAid™ Second Edition and getting started with SimUaid™". Brooks/Cole-Thomson Learning, Canada, 2002.

AUTHOR INDEX

Z